Forensic Electrical Engineering & Liability

Robert E. Nabours, Ph.D., P.E.

Lawyers & Judges Publishing Co.

Lawyers & Judges Publishing Co.
P.O. Box 30040
Tucson, AZ 85751-0040
(602) 751-1500
FAX - (602) 751-1202

Nabours, Robert E.
Forensic electrical engineering & liability / Robert E. Nabours.
p. cm.
Includes bibliographical references and index.
ISBN 0-913875-07-4 (hard)
1. Electric engineering—United States. 2. Forensic engineering—United States. 3. Evidence, Expert—United States. I. Title.
II. Title: Forensic electrical engineering and liability.
TA153.N27 1994
621.319'24—dc20 94-32203
CIP

Printed in the United States of America
10 9 8 7 6 5 4 3 2 1

Cover Illustration: Mr. Ouch logo © courtesy of NEMA.

Dedication

To my father, who gave me a devotion for reading and a love of nature.

Acknowledgements

The author is grateful to his friends and professional associates, Mike Beale and Bob Witter, for their careful reading and considerable attention to technical detail that has contributed to the accuracy and expression of ideas within this book.

Michael A. Beale is a practicing trial attorney with the law firm of Jennings, Strouss & Salmon, P.L.C. in Phoenix, Arizona.

Robert E. Witter is a professional engineer and practicing forensic electrical engineer with Robert Witter & Associates in Albuquerque, New Mexico.

Table of Contents

Preface

This book is intended to serve two distinct professionals: attorneys and electrical engineers. Neither profession will find the material presented in their specialty to be advanced in scope or depth of most technical matter.

Because attorneys and forensic electrical engineers must occasionally work together to serve within the judicial system, there are elements of each profession that the other must understand and accommodate for the symbiotic relationship to be useful.

The book explains what a forensic electrical engineer does and how this function is performed. The terminology of law, legal and voluntary standards and electrical codes and regulations are presented to inform electrical engineers of their role in the symbiosis. This material is covered in the first three chapters of the book. Practicing trial lawyers will also find that this material is useful in the perspective it affords regarding forensic electrical engineering and the law.

In Chapter Four, the attorney is introduced to a few basic concepts of electrical engineering. The electrical/hydraulic analogy, Ohm's law and basic circuits, alternating current and common modern power systems, wiring methods, materials and equipment are discussed in the terminology of the electrical engineer. The technical and theoretical content of this introduction should be understandable by a high school graduate who passed algebra, geometry, chemistry and physics. The intended readers of this book are at least this qualified in these subjects.

Chapter Five addresses the subject of how electrical energy affects human health. There are direct and indirect effects produced by man-made electrical energy on all life forms on earth and some of the more subtle phenomena are just beginning to be understood. Today's world is very different in its electromagnetic environment from that of fifty years ago, and mankind's welfare in this environment is being threatened. Many of the scientific facts, conclusions, and opinions expressed in this Chapter will be painful or judged unacceptable by some in the electrical industry; however, these facts, conclusions and opinions are delivered with the sincere intent to promote the safe service of electrical power to the public.

The last chapter of this text is a compendium of a selection of twelve of the author's experiences over a 25 year period as a forensic electrical engineer. Many of the over 300 cases which the author investigated did not result in litigation and of those which did involve the filing of a law suit, only about 16% actually resulted in a trial wherein the author testified. A far greater number of litigations were settled out of court before a trial began. Whether an engineering investigation eventually was utilized by an attorney to settle or try a law suit, the principles of forensic electrical engineering were applied in every case. Hopefully this book will accomplish the avowed purpose and both attorneys and electrical engineers will benefit from this view of a highly interesting and challenging professional specialty—forensic electrical engineering.

Chapter 1

The Forensic Electrical Engineer

The Engineering Expert Witness

A forensic engineer has several professional roles to fulfill. To this end the forensic electrical engineer must:

1) exercise a high level of technical competence in electrical engineering,
2) have the ability to serve the client as a technical consultant and specifically to be capable as an expert witness,
3) possess and demonstrate the necessary professional qualifications to testify as an expert in court.

Not all electrical engineers can or want to provide these essential functions of a forensic electrical engineer.

The procedures of the U.S. courts are adversarial in nature and many engineers choose to avoid conflict, particularly when their judgement or technical ability is called into question by an opposing litigant or another expert.

It is also not sufficient for an engineer to be both technically competent and willing to testify. The court must first qualify the engineer as an expert. In most judicial proceedings an expert must have "special knowledge, skill, experience, training or education relating to the nature of the case"[1] and it is the providence of the judge to rule on an expert's qualifications. Although an engineer is not required by law to hold a professional license in order to testify as an expert, in practice this is often a minimum requirement. First of all, most states require an engineering license before an engineer can provide services to the public. Secondly, the professional status afforded the engineer who has established his technical ability through professional licensing is more likely to produce credibility and weight of testimony in the courtroom.

The National Society of Professional Engineers (NSPE) has undertaken to study the specialty of Forensic Engineering and in 1980 published Guidelines for Forensic Engineering. Shortly thereafter, in 1982, three prominent professional engineers (PEs) formed the National Academy of Forensic Engineers* (NAFE)[2].

The NAFE seeks to improve the practice, elevate the standards, and advance the cause of forensic engineering. Membership in the Academy is limited to Registered Professional Engineers who are members of NSPE and also members in an acceptable grade of a recognized major technical engineering society[3]. For electrical engineers this would be the Institute of Electrical and Electronics Engineers (IEEE). Candidates for membership in the NAFE are required to demonstrate actual experience in forensic engineering, including testimony under oath subject to cross-examination. Recommendations are required from practicing attorneys or senior claims managers familiar with the candidate's forensic experience. In furtherance of its goals, NAFE holds meetings and conducts seminars at the biannual meetings of the NSPE which are held throughout the U.S. The NAFE publishes two issues of the Journal each year which include learned papers presented at the meetings and seminars.

Since the Journal of the NAFE was first published in 1984, there have been 119 articles published on the technical aspects of forensic engineering and 47 articles dealing with the general practice and ethical considerations of the profession. The NAFE membership is currently over 400 and with correspondents, international affiliates, and honorary designations, the organization's participation is nearly 500 individuals.

The Forensic Electrical Engineer's Clients

In the author's experience, over a 25 year period of practice, the first clients and by far the greatest number of clients have been attorneys. At first these were largely plaintiffs' attorneys, but as time progressed the mix between plaintiffs' cases and defense cases

---

* NAFE is a Chartered Affinity Group of NSPE with headquarters at 174 Brady Avenue, Hawthorne, NY 10532.

has become about equal. In most of these cases, the attorney representing one of the parties in a lawsuit directly contacted this expert. The initial contact by an attorney will usually explore the possibility of any conflict of interest in the case or between the parties involved and determine the willingness and capability of the expert to handle the assignment.

Plaintiffs' attorneys are initially interested in determining whether in the expert's opinion liability exists on the part of a potential defendant or defendants. A defendant's attorney may seek the expert's help in answering these same questions but, because a lawsuit has usually already been filed, will wish to explore the existence and gathering of possible evidence that will be beneficial to his client as a defense.

The ensuing preliminary investigation and report by the electrical engineering expert must be thoroughly and objectively prepared. The result of this preliminary work will usually serve to tell the attorney (client) whether a) they have a case, b) they have a case but with considerable technical difficulty, or c) they have no case at all.[4]

The actual parties in these litigations or potential lawsuits are plaintiffs who typically believe they have suffered injury by electrical product failure, contact with electrical energy, fire of electrical origin, failures of electrical systems, professional malpractice, or inadequate illumination. The defendants in such charges of liability include manufacturers, design professionals, contractors and installers of electrical equipment or systems, electric and communication utilities, transportation companies, owners of structures, owners of industrial plants and facilities, government agencies, and insurance companies.

Insurance companies are the second largest class of clients in the author's practice. Ordinarily a direct contact from an insurance company originates from a senior claims adjustor and the most frequent question to be answered is: "Did electricity cause the accident?" This includes fire, product failure and personal injury. The insurance company may be interested in the possible subrogation of a claim, or it may anticipate a law suit against one of its insured and, therefore, the expert is retained even before a lawyer is contacted. Fires of suspected electrical origin are the most numerous type of case directly referred to the author by insurance companies.

A third and fourth ranking of clients by number of cases referred would be the courts and individuals. The court may appoint an expert to serve as a "friend of the court" in which case the expert is not a witness for either plaintiff or defendant. The expert may be paid jointly and equally by the parties to the lawsuit in such cases.

On occasion the author has been retained directly by an individual to conduct a preliminary investigation regarding the efficacy of a potential litigation. Certain individuals or parties apparently prefer to examine the engineering electrotechnical aspects of a liability question before they retain a lawyer.

The Forensic Electrical Engineer's Function

Preliminary Investigation

As has already been stated, the first work of the forensic electrical engineer is to conduct a preliminary investigation of the case. This investigation may include a review of all fact witness statements or depositions, examination of photographs and evidence, a possible visitation to the site of the incident, a review of other experts' reports and depositions, and research for applicable codes and standards. If the client has only recently begun to assemble a file on the case, there may be little or none of the material cited above available to review.

It is frequently necessary in early discovery for the expert to suggest how and what material needs to be gathered in order to prepare a thorough analysis of the incident. Occasionally the initial information regarding a potential litigation should be recognized by the expert as decisively negative and it is incumbent upon the forensic engineer to be forthright in expressing this opinion.

Detailed Technical Investigation

Following a preliminary investigation, resulting in a realistic basis for proceeding with the litigation, the detailed technical work of case preparation begins. It is now the expert's job to recommend and conduct any mathematical analyses, physical experiments or field investigations that will assist the client in solidifying their case. It is also necessary for an objective technical investigation to anticipate the preparation of the other attorneys' experts and to prepare to accept or refute other technical explanations for the incident.

There are scientific literature databases available for computer searches both online and through university and general libraries that can greatly enhance the investigator's preparation of opinions and rebuttal of opposition opinion. These reference sources are extremely useful in complicated technical cases involving complex scientific issues.

An important part of an expert's job is to recognize their own limitations in knowledge and training that may need supplementation by other experts. There should be no disgrace brought upon an expert who says "I don't know the answer" or "That is beyond my area of expertise."

Comprehensiveness (*i.e.,* the extensive understanding of many things) is a valuable asset for a consulting forensic engineer. However, an expert's ego has no place in the court room. A NAFE engineer expressed it very well when he wrote: "The difference between a comprehensivist and a specialist is that the comprehensivist has a practical working knowledge of many areas of engineering and knows enough about several specialties outside of his own that he can recognize the need for other specialists and work with them."[5]

During the conduct of field investigations, experiments, and evidence examination, photographs and video tapes are valuable adjuncts to the forensic engineer's notes and observations. The ability to fully utilize these photographic tools for supplementing the investigation is a good reason for comprehensiveness by the expert.

As the technical investigation progresses, the attorney will obtain additional information from sources not directly available to the engineering expert. A close working relationship with the attorney will allow the expert to be aware of these information sources and to obtain copies of relevant materials. Examples of such information are police reports, fire and insurance investigation reports, weather reports, subpoenaed manufacturer's data, depositions, and answers to interrogatories.

Electrical cases frequently require a hypothetical explanation. The forensic electrical engineer, therefore, must utilize the scientific method for hypothesis testing in order to present a quantum of proof to the court that weighs heavily in favor of the proposed explanation for the incident. This has been found to be particularly true of electrical product failure and fire cases.

All possible hypotheses must be considered and the least likely eliminated by objective scientific and mathematical analysis. The most likely postulated explanation for the incident will require a preponderance of rigorous proof that can be explained to the court and made understandable to the jury. The reduction of frequently complicated mathematical and scientific facts to a format that can be understood by the layman is a challenge for an engineer, even one who has experience in teaching beginning engineering students.

Final Report

After the available facts have all been ascertained and the technical investigation completed, a final report can be prepared by the expert. The report should include references to all sources relied upon in reaching the conclusions and opinions expressed by the forensic engineer. The report should be discussed between the attorney and the engineer prior to committing the work in writing. Frequently the attorney will request that the engineer avoid a written final report since it will then be subject to discovery by the other litigants. Most material in the engineer's case file is discoverable by all parties to the law suit, and all notes, calculations, or other work product by the engineer should be prepared with such ultimate disclosure in mind.

If the final report and expert's conclusions are favorable to the attorney's case, the expert will usually then be identified and disclosed to the other litigants and, in some cases, the court may compell such disclosure. Should the results of the engineer's investigation not support the attorney's case, the engineer may or may not be listed or called upon to testify at trial. Whether or not the engineer is named as a witness, the results of the engineer's work on the case should be treated with confidentiality and should not be revealed to anyone except the client or his attorney unless asked by opposing counsel or ordered to do so by the court or client attorney. If required by subpoena to testify in deposition or at trial, the engineer must, of course, answer fully and truthfully all questions posed to the best of his/her ability. Frequently, in the author's experience, the technical investigation produces very strong evidence that may be used to persuade the other litigants to settle the case out of court. Occasionally, the facts discovered about the alleged liability of a party have been so persuasive that the attorney has requested the author to prepare an affidavit to support a motion for summary judgement requesting dismissal of the lawsuit by the court. This process can be used,of course, in the reverse to resist efforts by the adverse party for summary judgement.

Discovery

Attorneys customarily use two processes to formally inquire into the opposing counsels' case: the interrogatory and the deposition. Interrogatories are written questions submitted by an attorney to the opposing attorney requesting written answers under oath by a litigant. The forensic engineer is frequently asked to help prepare technical questions for the interrogatories. Such questions are discussed by the engineer and the attorney to assist the lawyer in understanding the technical aspects of the case and to properly word the questions in order to increase the likelihood of receiving a useful response. These technical questions may be referred to the engineer for assistance in preparing the answers. Again, the engineer and attorney will discuss the questions or answers to improve the lawyer's understanding of the details of the case and to avoid answers that do not specifically reply to the question.

A deposition is testimony taken under oath outside the court. At a deposition the questions from an attorney to a witness and the answers thereto are taken down by a court reporter who then produces a written verbatim record. An expert is notified by subpoena or by informal arrangement through the office of the client attorney that his/her deposition will be taken at a certain time. On that date the expert, client attorney, opposing attorney and court reporter meet to conduct the deposition. There may be several attorneys and sometimes, when the lawyers are in agreement, other experts present during a deposition.

A deposition is the equivalent of court testimony and all statements taken by deposition can be ruled admissible by the judge during a trial. The intended purpose of a deposition is for the opposing attorney to discover what the witness will testify to at trial. In reality depositions are used for many other purposes including: an exploration of the other attorney's case, an appraisal of the expert's capability and credibility as a witness, and an attempt to create ambiguity in the expert's testimony that can later be used to discredit the witness before a jury.

Frequently, in the author's experience about 75% of the time, a deposition is taken but the case never goes to trial. Sometimes this is due to the strength or weakness of the case which is exposed during the deposition leading to a settlement out of court.

In other instances the expert's deposition is taken by the opposition attorney to convince their client that the case is unlikely to be won in court.

Giving a deposition is a critical point in the expert's preparation of a case. The art of testifying in deposition has been explored by forensic engineers[6, 7] and attorneys[8] in numerous publications and is too lengthy a subject to pursue adequately in this text.

Testifying at Trial

In order to prepare for testimony in court, the expert should review thoroughly all work done during the technical investigation, all depositions taken by other experts that bear upon the planned testimony, and all reports and depositions given by the engineer. The attorney and engineer should meet prior to trial to discuss the planned testimony and to be sure that both anticipate the opposing party's examination. It is also important to discuss the opposing attorney's strategy for trial as far as known. The client attorney may have prior experience with the other attorney that can be helpfully shared with the expert.

During the technical investigation and at the conclusion of the discovery process, the engineer and attorney should have decided what exhibits would be useful at trial. Certain graphs, charts, or blowups of mathematical calculations can be prepared by or at the direction of the engineer. Photographic enlargements and physical evidence are also frequently offered as exhibits. The engineer should study all exhibits prior to his/her court testimony, both those offered by the client and those of the opposition.

The direct examination is usually straight forward with questions asked of the expert by the client attorney. The attorney will review the expert's qualifications including education, training, and experience. Then the forensic engineer will be asked to explain the investigation, theories, experiments, and analyses of the technical aspects of the case. Finally the expert will be asked for his/her conclusions and opinions based upon these facts.

The direct examination may include questions regarding the work of other experts and the engineer's opinions of the validity of that work. A professional, objective, and impersonal critique of the other expert's work is often helpful in convincing the jury of an engineer's credibility and honesty.

Following the direct examination, the other attorney will conduct a cross examination. The cross examination can be a stressful experience for both the engineer and the attorney. Very few attorneys will be as technically qualified as the expert in their field, and questions posed indiscriminately of an expert can produce answers that are ruinous to the attorney's case. On the other hand, it is the attorney's job to offer as strong an opposition to an expert's damaging opinions as can be found and most qualified attorneys are prepared for this exercise by their scientific consultants.

In some cases the attorney will directly attack an expert's qualifications or attempt to show bias by the expert in favor of the expert's client or case. If an expert has a conscious or subconscious bias in favor of his client, a competent cross examination will undoubtedly make this clear. An engineer may have a justifiable bias toward their own work and opinions, but they cannot serve effectively as a forensic expert if they are biased toward the client or case.[9]

Other subjects for the cross examination of an expert are errors in assumptions, incorrect facts, judgement errors, inconsistent testimony in prior cases, and errors in engineering technology or science. *Note: All of these targets for cross examination can be essentially eliminated by the expert through careful preparation of the technical work on the case.*

Certain facts may be given by fact witnesses, and these can be taken by the engineer at face value and often used as a predicate for hypothetical opinions. Such facts should be given a credibility in proportion to their consistency with other witness observations, their agreement with the physical laws of nature, and the perceived bias or truthfulness of the witness. Other facts are developed by the technical investigation and these are, and generally should be, subject to analytic verification in most cases.

Errors in engineering technology or science can best be avoided by checking and rechecking the work. Often a review by a colleague or another expert with a different specialty will be appropriate when the analytical or scientific work is complicated or critically dependent upon assumptions.

As was the case with testifying in deposition, testifying under cross examination is a subject with too much detail to explore further in this text.

Following cross examination the client attorney may ask additional questions of the expert under redirect examination. Only the subject matter brought out under cross examination is permitted to be addressed under redirect. The attorney can use redirect to allow the expert to expand or clarify answers given under cross examination, however, a good cross examination will rarely give the expert this opportunity. This is an excellent opportunity (and offers the only one) for the engineer to clarify brief answers given to convoluted hypothetical questions posed by the other attorney and to explain unrealistic assumptions contained therein.

As a matter of policy, the author retains a permanent file of all original work on forensic cases. Material which was furnished by others and can reasonably be expected to exist elsewhere is purged from the file after sufficient time has transpired to be assured that the case has not been appealed.

Chapter Notes

1. D.G. Sunar, Ph.D. *The Expert Witness Handbook—A Guide for Engineers,* Professional Publications, Inc., 1985. p. 2.

2. Marvin M. Specter, P.E. "Recollections of Ten Years of NAFE," *Journal of the NAFE,* Vol. IX, No. 2, December 1992. pp. 1 to 20.

3. *NAFE Directory,* 2nd ed., the NAFE, 1991. Introduction.

4. Marvin M. Specter, P.E. "The Benefit of Negative Opinion," *Journal of the NAFE,* Vol. III, No. 2, 1986. p. 15.

5. Bill E. Forney, P.E. "The Forensic Engineer as the Comprehensivist," *Journal of the NAFE,* Vol. III, No. 2, 1986. p. 87.

6. D.G. Sunar, Ph.D. *The Expert Witness Handbook—A Guide for Engineers,* Professional Publications, Inc., 1985. pp. 18 to 27.

7. Robert L. Bleyl, Ph.D., P.E. "The Forensic Engineer on the Witness Stand Under Cross-Examination Attack," *Journal of the NAFE,* Vol. II, No. 2, December 1985. pp. 15 to 19.

8. Thomas A. Bratten, J.D., P.E. "The Forensic Engineer on Cross- Examination," *Journal of the NAFE,* Vol. IV, No. 1, June 1987. pp. 67-78.

9. ibid. p. 70.

Chapter 2

Standards and the Law

Terminology

Electrical codes and standards will be presented in detail in Chapter Three. Throughout that chapter there will be terminology used that needs definition to assure that a precise meaning is conveyed. The following terms are defined in the context of standardization:

Standard—commonly used and accepted as an authority;[1] a document setting forth requirements normally dictated by customary practices in industry, science, or technology.[2]

Code—a systematic collection of regulations and rules of conduct or procedure;[1] a body of recommendations of good practice to satisfy certain considerations (*e.g.*, safety, quality, economy, or performance).[2]

Recommended Practice—a standards document that provides information on good engineering practice;[2] advised or counseled procedure for conducting an engineering activity (*e.g.*, design, analysis, or synthesis).

Guides—a standards document that provides alternative information which comprises good engineering practice.[2]

National Standard—a standard that has been adopted by a recognized national standards body, such as the American National Standards Institute (ANSI);[2] a standard that is recognized and used nationally by preference.[2]

Legality and Liability

The question of legality of a standard can be separated basically into two considerations. First, is the standard acceptable under U.S.

laws relating to price fixing and trade restraints? The Federal Trade Commission and Department of Justice are concerned with this aspect of the creation or adoption of standards. A national standard such as that adopted by ANSI is extremely unlikely to violate these precepts. Standards created by participants who follow an unprejudiced and fair development practice leading to an equitable resolution of the economic, safety, quality, and performance requirements for a product, process or procedure are also unlikely to have their work questioned legally.

The second consideration in legality of a standard has to do with the question of conformance under the law by a party accused of an act which is alleged to violate the standard. Unless the standard has been enacted into law by a governmental body having jurisdiction in the matter, it is not a mandatory legal requirement, the breach of which imposes liability, but rather it may and frequently does become a measure by which the jury may evaluate "good practice." When the standard does not strictly govern the behavior of a party, the question may revert to a jury who may interpret the nonconformance with a recognized and customary practice in the industry (*i.e.,* a standard) to be sufficient cause for liability.

Certainly in most cases standards are essential to the law and, as previously discussed, even when a standard does not carry the force of statutory liability in a litigation, the violation of a standard may and often does persuade a jury to find for the plaintiff. Standards may also be a strong defense for the defendant, and the compliance with both adopted standards and non-adopted but widely accepted standards can frequently resolve a liability issue in favor of the party in compliance, often without the necessity and expense of a trial. Today, the National Electrical Safety Code (NESC) and the National Electrical Code (NEC) are viewed by most U.S. Courts as dispositive so that most legal-electric issues and juries therein are frequently instructed accordingly.

A Brief History of Electrical Standards

Throughout this text standards have been treated in the legal sense. We have focused on the use of electrical standards for the establishment of a norm of conduct under the law.

Standards are essential in any scientific endeavor for the avoidance of conflict and needless duplication of research. A new technology first develops the standards necessary to scientific measurement and communication of research and study. Later, as the technology is applied by engineers, the need for safety standards, codes, specifications and laws becomes essential to the practical use of the technology. The development of standards in electrical science began in the U.S. around 1880. There were early problems to be resolved regarding basic definitions and measurements for such phenomena as self-induction, resistance of conductors and electromotive force. It was not until 1896 that a first conference on standard electrical rules was held which produced the "Underwriter's Rules" that eventually led to the National Electrical Code.

In 1897 the standard of luminous intensity (candlepower) was adopted. Then in 1898 and 1899 the first American Institiute of Electrical Engineers (AIEE) product standards committees were formed to address the subjects of generators, motors and transformers.

In 1901 the U.S. Congress established a national bureau for standardization which became known as the National Bureau of Standards (NBS) which greatly affected the growth of electrical technology in the U.S. Progressively the process of standardization moved from basic scientific standards to technical standards and in about 1920, following World War I, into manufacturing standardization.

International Electrical Standards

International electrical standards were begun in 1904 at St. Louis and have now evolved into the International Electrotechnical Commission (IEC). The International Electrotechnical Commission (IEC) has membership comprised of National Committees from 42 nations. The technical work of the IEC is done in Technical Committees (TCs). Currently there are 78 TCs, many of which are subdivided into subcommittees.

The generation of IEC Standards begins in a Working Group created by a TC. Leading experts in the field of work are selected from the individuals offered by the National Committees. These experts participate in the Working Group as individuals, not as representatives of their National Committees.

The Working Group develops a draft for a proposed standard which is forwarded to the IEC central office in Geneva and circulated to all National Committees. Comments on the document may be referred back to the Working Group, and the process repeated until the TC decides that a draft is ready for final ballot. A 20% negative vote by the National Committees will defeat adoption of the proposed standard. Eventually, if a draft has been accepted, after all editorial changes are resolved and another ballot taken, the document becomes an IEC Standard.

The U.S. participates intimately in the IEC through the U.S. National Committee (USNC). The USNC is now a part of ANSI and is run by an Executive Committee of elected officers with an ANSI staff member acting as secretary. Each member of the Executive Committee assumes responsibility for a selected number of IEC TCs through a group of individuals called Technical Advisers (TAs). The TA is responsible for organizing a committee of experts, usually an existing committee in the U.S. This can be an existing ANSI committee or a committee from some other organization such as an engineering society, manufacturer's organization, or a special committee appointed for the specific TC area.

The Commission Internationale de l'Eclairage (CIE) was formed in 1913 to deal with illumination. The Illuminating Engineering Society of North America (IESNA) is a member of the CIE and has worked closely with the international organization to standardize and improve the art and science of illumination.

The CIE has approved a system that can identify and designate color, tint and shade by instrumentation, eliminating visual judgment. The standard instrument used to measure color or more specifically the radiant power at various wavelengths is called a spectrophotometer.[3] This system has resulted in a specification for light sources which allows their color rendering properties to be compared analytically.

Among other standards developed by or approved by the CIE are luminaire classifications, spectral luminous efficiency values and standard illuminants.

Voluntary Standards

Standards that are developed by the voluntary activities of professional societies (*e.g.*, the Institute of Electrical & Electronic Engineers [IEEE]), trade associations (*e.g.*, the National Electrical Manufacturer's Association [NEMA]), or other organizations (*e.g.*, Underwriters Laboratories [UL]), are effectively the approved recommendations of these organizations. Unless these standards are adopted by governmental bodies as regulatory provisions or incorporated into laws and statutes by Congress, state legislatures, or local governments, they have no mandatory legal force.*

Conformance to a standard can neither be presumed nor is conformance mandatory unless prescribed by law. Many standards are developed by manufacturers for reasons other than safety of the consumer or the public in general. Trade association standards are frequently adopted to represent the technical quality of products for warranty limitation purposes.

The effectiveness of the voluntary standards system in the U.S. is largely due to the activities of ANSI. Those standards which are independently developed by multiple organizations are not adopted as national standards by ANSI without a thorough review of the various proposals for technical accuracy, fairness and conflict. Only standards that are technologically sound and for which the need can be established become ANSI designated national standards.

Chapter Notes

1. Webster's II. *New Riverside University Dictionary,* Riverside Publishing Company, 1988.
2. Fink and Beaty, eds. *Standard Handbook for Electrical Engineers,* 13th ed., McGraw-Hill, 1993. pp. 28-6 to 28-8.
3. Louis, Erhardt. *Radiation Light and Illumination,* Camarillo Reproduction Center, 1977. pp. 70 to 76.

---

* See *Legality and Liability,* in this chapter, pp. 21–22.

Chapter 3

Electrical Codes and Standards

American National Standards Institute (ANSI)

ANSI* coordinates and promotes the development of standards and approves as American National Standards documents that have been prepared in accordance with its regulations. ANSI evolved from the American Engineering Standards Committee (AESC) which was formed in 1919 as a result of development by the American Institute of Electrical Engineers (AIEE—founded in 1884). The AIEE has since merged with the Institute of Radio Engineers (IRE—founded in 1912) to form the Institute of Electrical and Electronics Engineers (IEEE—formed in 1963).

Standards that have been developed and approved by other organizations may be approved by ANSI, in which case they carry the identification numbers of both organizations. Such standards are available from either organization. ANSI also appoints committees for standards development that are organized and administered by other organizations. Generally such committees are formed to coordinate the work of a large number of organizations. Standards developed by ANSI committees will carry only an ANSI number. ANSI standards of interest to forensic electrical engineers include the following:

1. ANSI/IEEE Std. 141-1993, IEEE Recommended Practice for Electric Power Distribution for Industrial Plants, IEEE Red Book.

---

* ANSI is headquartered at 1430 E. Broadway, New York, NY 10018.

2. ANSI/IEEE Std. 142-1991, IEEE Recommended Practice for Grounding of Industrial and Commercial Power Systems, IEEE Green Book.
3. ANSI/IEEE Std. 241-1990, IEEE Recommended Practice for Electrical Power Systems in Commercial Buildings, IEEE Gray Book.
4. ANSI/IEEE Std. 242-1986, IEEE Recommended Practice for Protection and Coordination of Industrial and Commercial Power Systems, IEEE Buff Book.
5. ANSI/IEEE Std. 399-1990, IEEE Recommended Practice for Power Systems Analysis, IEEE Brown Book.
6. ANSI/IEEE Std. 446-1987, IEEE Recommended Practice for Emergency and Standby Power Systems for Industrial and Commercial Applications, IEEE Orange Book.
7. ANSI/IEEE Std. 493-1990, IEEE Recommended Practice for the Design of Reliable Industrial and Commercial Power Systems, IEEE Gold Book.
8. ANSI/IEEE Std. 602-1986, IEEE Recommended Practice for Electric Systems in Health Care Facilities, IEEE White Book.
9. ANSI/IEEE Std. 739-1984, IEEE Recommended Practice for Energy Conservation and Cost-Effective Planning in Industrial Facilities, IEEE Bronze Book.
10. ANSI/IEEE Std. 1100-1992, IEEE Recommended Practice for Powering and Grounding Sensitive Electronic Equipment, IEEE Emerald Book.
11. ANSI/IEEE Std. 100-1984, IEEE Standard Dictionary of Electrical and Electronic Terms.
12. ANSI/NFPA 70-1993, National Electrical Code.
13. ANSI C2-1993, National Electrical Safety Code.
14. ANSI/IEEE Std. C95.1-1991, Standard Safety Levels with Respect to Human Exposure to Radio Frequency Electromagnetic Fields, 3 kHz to 300 GHz.

National Electrical Code (NEC and NFPA)

The National Fire Protection Association (NFPA)* has acted as sponsor for the NEC since 1911. The NEC is periodically reviewed and revised to reflect changes in both technology and hazard recognition within the electrical industry. The NEC Committee includes representatives from all phases of electrical technology (*e.g.*, manufacturers, contractors, electrical building inspectors, utilities, engineering organizations, fire underwriters, and independent testing laboratories).

The NEC has been published since 1897 with periodic superseding editions. The 1993 edition is the latest edition, with a 1996 revision currently in process. The purpose of the NEC is the practical safeguarding of persons and property from hazards arising from the use of electricity.

The NEC covers:[1]

1. installations of electric conductors and equipment for public and private buildings and structures;
2. installations of conductors and equipment that connect to the supply of electricity (utility service);
3. installations of other outside conductors and equipment on the premises; and
4. installations of optical fiber cable.

The NEC does not cover:[1]

1. installations in ships, watercraft other than floating buildings, railway rolling stock, aircraft, or automotive vehicles other than mobile homes and recreational vehicles;
2. installations underground in mines and self-propelled mobile surface mining machinery and its attendant electrical trailing cable;

* NFPA is located at Batterymarch Park, Quincy, MA 02269.

3. installations of railways for generation, transformation, transmission, or distribution of power used exclusively for operation of rolling stock or installations used exclusively for signaling and communications purposes;
4. installations of communications equipment under the exclusive control of communications utilities located outdoors or in building spaces used exclusively for such installations; or
5. installations under the exclusive control of electric utilities for the purpose of communications or metering; or for the generation, control, transformation, and distribution of electric energy located in buildings used exclusively by utilities for such purposes or located outdoors on property owned or leased by the utility; or on public highways, streets, roads, etc., or outdoors by established rights on private property.

The intent of the NEC is to cover all premises wiring or wiring other than utility-owned metering equipment, on the load side of the service point of buildings, structures, or any other premises not owned or leased by the utility. Also it is the intent to cover installations in buildings used by the utility for purposes other than listed above, such as office buildings, warehouses, garages, machine shops, and recreational buildings that are not an integral part of a generating plant, substation, or control center.[2]

The NFPA publishes extensively on the subject of fire safety. Besides the NEC, there are other NFPA products of interest to electrical engineers including:

1. The NEC Handbook, Edited by Mark W. Earley, Richard H. Murray and John M. Caloggero.
2. Electrical Installations in Hazardous Locations, Peter J. Schram and Mark W. Earley.
3. American Electricians' Handbook, Terrell Croft and Wilford J. Summers, McGraw-Hill.

There are also NFPA Codes and Standards published in pamphlet editions:

4. NFPA 70A, "Electrical Code for One- and Two-Family Dwellings and Mobile Homes."
5. NFPA 70B, "Electrical Equipment Maintenance."
6. NFPA 70E, "Electrical Safety Requirements for Employee Workplaces."
7. NFPA 72, "National Fire Alarm Code."
8. NFPA 73, "Residential Electrical Maintenance Code for One- and Two-Family Dwellings."
9. NFPA 75, "Protection of Electronic Computer/Data Processing Equipment."
10. NFPA 77, "Static Electricity."
11. NFPA 79, "Electrical Standard for Industrial Machinery."
12. NFPA 110, "Emergency and Standby Power Systems."
13. NFPA 111, "Stored Energy Emergency and Standby Power Systems."
14. NFPA 262, "Method of Test for Fire and Smoke Characteristics of Wires and Cables."
15. NFPA 297, "Telecommunications Systems-Principles and Practices for Rural and Forestry Fire Service."
16. NFPA 496, "Purged and Pressurized Enclosures for Electrical Equipment."
17. NFPA 497A, "Classification of Class I Hazardous (Classified) Locations for Electrical Installations in Chemical Process Areas."
18. NFPA 497B, "Recommended Practice for the Classification of Class II Hazardous (Classified) Locations for Electrical Installations in Chemical Process Areas."
19. NFPA 497M, "Classification of Gases, Vapors and Dusts for Electrical Equipment in Hazardous (Classified) Locations."
20. NFPA 780, "Lightning Protection Code."

21. NFPA 850, "Electric Generating Plants."
22. NFPA 851, "Hydroelectric Generating Plants."
23. NFPA 907M, "Manual for the Determination of Electrical Fire Causes."

The NFPA also publishes Special Topics References:

24. Item No. JP-RES-27, "Electrical and Instrumentation Safety for Chemical Processes."
25. Item No. JP-RES-28, "Electrical Hazards and Accidents."

National Electrical Safety Code (ANSI and IEEE)

The National Electrical Safety Code (NESC), ANSI C2-1993, is developed by American National Standards Committee C2 and published by the IEEE.* The NESC work began in 1913 with the publication by the National Bureau of Standards of NBS Circular 49 and was continued by NBS until H81, American Standard C2.2-1960 was issued in November 1961. Work was again commenced in 1970 by the ANSI C2 Committee resulting in the publication by the IEEE of ANSI C2- 1973.† Six updated editions of the NESC have been published since 1973 by the IEEE.

The purpose of the NESC is the practical safeguarding of persons (both utility employees and the public) during the installation, operation, or maintenance of electric supply and communication lines and associated equipment.

The NESC rules cover supply and communication lines, equipment, and associated work practices employed by an electric supply, communication, railway, or similar utility in the exercise of its function as a utility. They cover similar systems under the control of qualified persons, such as those associated with an industrial complex or utility interactive system.

---

* IEEE headquarters are at 345 East 47th St., New York, NY 10017.

† Ref. NESC C2-1990, Corrected Edition, 3rd Printing. Forward.

The NESC covers utility facilities and functions from the point of origin of the utility's supply of electrical energy or communication signals through the utility system to the point of delivery to a customer's facilities. In the case of an electric supply utility, this is the point of metering the customer's electric service. For a communications utility this is the customer's antenna, receiver, or line termination point.

NESC rules do not cover installations in mines, ships, railway rolling equipment, aircraft, or automotive equipment, or utilization wiring except as necessary for the installation and maintenance of electric supply stations and underground electric supply and communication lines as covered by Parts 1 and 3 of the Code.[3]

The membership of ANSI Committee C2 includes representatives from the utilities, insurance groups, railroads, public transit association, National Safety Council, electrical inspectors, area power administrations, and the Canadian Standards Association.

The IEEE publishes Interpretations of the NESC prepared by a representative committee acting upon requests for interpretations of the rules contained in the Code. Five volumes of Interpretations have been published to date, beginning with the period from 1961-1977 inclusive and concluding with the 1988-1990 inclusive period. Interim editions of Interpretations are collected and published on a regular basis by the IEEE.

IEEE Standards

The IEEE publishes numerous standards relating to the various fields in electrical and electronics engineering. Many of these standards are recognized and designated as ANSI Standards. Some have also been approved by the Canadian Standards Association and are CSA designated. Besides the ANSI/IEEE Standards cited in the ANSI section of this text, the IEEE publishes standards in areas of aerospace and electronics, broadcasting and communications, computers, power engineering, instrumentation and measurement, insulators and insulation, magnetics, motors and generators, nuclear power, recording, symbols and units, and electrical transmission and distribution.

IEEE standards are developed through ballot within the technical committees, and final approval is required by a Standards Board consisting of leading experts in a broad range of electrotechnology. The IEEE is the largest engineering society in the world (over 250,000 members). Membership is available to qualified individuals in engineering and science. No company memberships are offered.

Electronic Industries Association (EIA)

The EIA* was originally formed in 1925 and since has absorbed several electronic industry associations. Standards are developed by committees which operate under divisional panels representing various segments of the industry: Consumer Electronics, Distributor Products, Communications, Industrial Electronics, Component Parts, Solid State Products, Electron Tubes, and Government Electronics.

Among EIA services are activities such as compiling statistics, developing standards which are usually submitted to ANSI, presenting consensus views and recommendations, and representing the industry in governmental agencies and public hearings. The EIA annually publishes several periodicals including a Consumer Electronics Annual Review and the EIA Standards and Engineering Publications.

National Electrical Manufacturers Association (NEMA)

NEMA† was formed in 1926 and has grown to a membership of more than 550 companies engaged in the manufacture of electrical products for such fields as construction, power electronics, industrial electrical equipment, insulation, lighting, and wire and cable.

NEMA technical committees are comprised of engineers who represent member companies such that product standards are developed that realistically take into consideration the economic and technological factors which are essential to practical standardization in a manufacturing industry. Standards are adopted by

---

* EIA is located at 2001 Eye St. NW, Washington, DC 20006.

† NEMA is located at 2101 L St. NW, Washington, DC 20037.

consensus of the membership with final approval by the NEMA Codes and Standards Committee.

NEMA generates four classifications of standards:

1. NEMA Standard—defines a commercial product subject to repetitive manufacture.
2. Suggested Standard for Future Design—suggests an approach to product improvement or development.
3. Authorized Engineering Information—explains data or information in conjunction with NEMA standards.
4. Official Standards Proposal—standards draft for adoption submitted to some other organization (*e.g.*, ANSI).

ANSI has appointed NEMA as the secretariat organization for a number of standards committees including:

1. Insulated Wire and Cable
2. Dry Cells and Batteries
3. Attachment Plugs and Receptacles
4. Low Voltage Fuses, rated for less than 600 Volts
5. Industrial Control Apparatus
6. Insulators for Electric Power Lines
7. Switchgear
8. Transformers, Regulators and Reactors
9. Capacitors
10. Surge Arresters
11. Preferred Voltage Ratings for AC Systems and Equipment

Underwriters Laboratories (UL)

Underwriters Laboratories (UL)* was founded in 1894. UL is an independent not-for-profit organization originally engaged in testing products for electrical and fire hazards for insurance companies. Today the scope of UL has expanded to include the investigation and testing of materials, devices, products, equipment, constructions, methods and systems for hazards to life and property.

Engineering Services of UL evaluates appliances, rebuilt motors and generators, and electrical construction materials and wiring used to distribute power inside buildings and structures (*i.e.*, electrical systems covered by the NEC). Approved (Listed) products are given a UL label or mark. UL also lists (publishes) a directory of companies who have demonstrated the ability to produce certain products conforming to its requirements. UL submits its standards to ANSI for adoption as national standards.

Government Regulatory Standards

CPSC

The Consumer Products Safety Commission (CPSC) was established by Congress in 1972. In connection with its primary function, which is the prevention of injuries associated with consumer products, the CPSC compiles a Consumer Product Hazard Index which ranks such products by severity and frequency of accidents. The CPSC publishes Regulations which result from the development of safety standards for various types of products, which in the electrical field include electric toys, TV sets and appliances.

Upon a determination of the need for a standard by CPSC, a notice is published in the Federal Register. Interested persons or organizations are invited to submit standards for consideration toward this need. After the CPSC has evaluated such submissions, the Commission will publish the accepted Standard as a Regulation in the Federal Register.

---

* UL headquarters are at 333 Pfingsten Rd., Northbrook, IL 60062.

FCC

The Federal Communications Commission (FCC) was begun in the late 1920s as the Federal Radio Commission (FRC) to deal with the problems that were being created by radio frequency interference and frequency overlaps occurring in a growing broadcasting industry. The FCC Act of 1934 converted the FRC into the Federal Communication Commission. Since 1934 the FCC has grown to become one of the government's most extensive bureaucratic agencies with sections for broadcasting, safety and special radio services, cable television, field operations and common carrier (telecommunications).

FCC Rules are developed after a convoluted process of initiation, evaluation, notices of inquiry or proposed rule making, docket issuance, comment and reply evaluation, report and orders, public petitions for reconsideration and modifications.[4] The Commission cooperates with many organizations including the IEEE, the EIA, ANSI, the International Radio Consultative Committee (CCIR), the Federal Telecommunications Standards Committee (FTSC), and the Law Enforcement Standards Laboratory (LESL).

NRC

The Nuclear Regulatory Commission (NRC) was established in 1974 when the Atomic Energy Commission functions of regulation, research and development were separated.

The NRC provides Regulatory Guides that specify how an applicant for an NRC license may satisfy the Commission's safety criteria. The Regulatory Guide is not mandatory as a result of its issuance. The NRC attempts to cooperate and work with a voluntary standards system developed by leading organizations involved in nuclear power standards such as ANSI, American Society of Mechanical Engineers (ASME), American Society of Testing Materials (ASTM), and IEEE.

In-house, the NRC develops basic standards that define a level of risk judged acceptable to the public; the NRC also develops standards on assumptions for evaluating the consequences of nuclear plant accidents.

OSHA

The Occupational Safety and Health Administration (OSHA) was established as a Division of the U.S. Department of Labor by

Congress in 1970. OSHA's primary function is the prevention of injuries in the workplace. Like the CPSC, OSHA utilizes the Federal Register to both solicit comments and publish proposed rules and, after multiple review and final approval, to publish the standard as an OSHA regulation.

Many of OSHA's regulations and rules are taken directly from such ANSI standards as the NEC and the NESC. OSHA has promulgated rules for the operations of cranes and other forms of high profile equipment in proximity to overhead power lines. Violation of OSHA regulations can, in many jurisdictions, result in the imposition of legal liability as discussed in Chapter 2.

Uniform Building Code (UBC)

The UBC* is intended to develop better building construction and greater public safety by uniformity in building laws.[5] The UBC was first enacted by the ICBO in 1927. The Code has been revised at approximately three-year intervals since first enacted.

The UBC cites Related Publications for user knowledge of code enforcement and administration of a building inspection program. Among the related publications cited by the UBC is the National Electrical Code (NEC) which the UBC describes as "the electrical code for the majority of states, counties and cities in the United States."

The ICBO, the International Association of Plumbing and Mechanical Officials (IAPMO), the International Association of Electrical Inspectors, the Pacific Coast Electrical Association and NEMA have promulgated an administrative document to enable the UBC to be used by code enforcement agencies throughout the nation. Electrical systems specified as minimum standard requirements under the UBC are generally covered by other standards (*e.g.*, the Uniform Fire Code and the NEC).

---

* The UBC is published by the International Conference of Building Officials (ICBO), 5360 S. Workman Mill Rd., Whittier, CA 90601.

State and Local Standards

Individual governmental bodies exercising legal jurisdiction over electrical installations and utilities as well as insurance underwriters may choose to adopt all, some or none of the national standards that have been cited herein; however, the legal implications for such actions are significant.

Predominantly the states and local governments in the U.S. have chosen to adopt the widely accepted standards as promulgated by ANSI, particularly the NEC and the NESC. Many states have their own codes for electrical work that are modeled after the NEC with certain exceptions, usually enacting more conservative or restrictive requirements. The State of California, for example, through the Public Utility Commission (PUC), has its own code for electric utilities. PUC General Order No. 95 covers overhead lines and G.O. No. 128 deals with underground facilities.

State regulations which are in conflict with parallel Federal regulations are preempted, and the courts will apply the applicable Federal regulations. However, the degree of a standard is not always clearly defined and frequently an engineering opinion is necessary to assist a court or judicial hearing in its interpretation of standards.

Most states have commissions or regulatory bodies responsible for the regulation of public utilities that operate within the state. Frequently, in the case of electric power and communication utilities, the state utility commissions adopt the NESC as the minimum standard of operation for the applicable utility.

Illumination Standards (IESNA)

The Illuminating Engineering Society of North America (IESNA)* was founded in 1906 for the purpose of disseminating knowledge relating to the art and science of illumination. A Lighting Handbook was first published by the IES in 1949. Since then the IES has periodically updated and expanded its Handbook to accommodate its research and application activities.

* The IESNA is headquartered at 345 E. 47th St., New York, NY 10017.

In 1926 the IES created a Committee on Roadway Lighting, publishing The Principles of Streetlighting in 1928. This committee's work continued and resulted in the American Standard Practice for Street and Highway Lighting, published in 1947. The first ANSI standard for Roadway Lighting was published in 1972 and has evolved to include the American National Standard Practice for Roadway Lighting, RP-8 1983 which is a part of the U.S. Department of Transportation, Federal Highway Administration's Roadway Lighting Handbook.

The purpose of the Standard Practice is to serve as a basis for design of fixed lighting for roadways, bikeways and pedestrian ways. It is not intended as, nor does it establish a legal standard for roadway lighting systems.[6] The adoption of the IES Standard Practice by governmental agencies, however, creates a legal standard, and certainly the principles of design recommended by the Standard Practice represent state-of-the-art engineering for lighting systems. At the very least the IES Standard Practice is an accepted standard for good lighting.

The IES Lighting Handbook is currently published in one volume with two sections, a reference section and an applications section. The Lighting Handbook treats the subject matter of terminology, physics of light, light and vision, measurements, luminaire design, daylighting, light sources, lighting calculations, lighting applications, and codes and regulations. Importantly, the IES has published recommendations for illuminance (density of incident light energy) for a broad classification of visual activities.

Chapter Notes

1. ANSI/NFPA 70, *National Electrical Code—1993,* NFPA. Art. 90-2, pp. 70-1, 2.
2. Ibid. (FPN), p.70-2.
3. ANSI C2-1990, *National Electrical Safety Code—1990,* IEEE. Sect. 1.011, Scope, p. 45.
4. Fink and Beaty, eds, *Standard Handbook for Electrical Engineers,* 11th ed., McGraw-Hill, 1978. pp. 29-17 to 29-20.
5. *Uniform Building Code 1991 ed.,* the International Conference of Building Officials. Part I, Ch. 1, Sec. 102, Purpose.
6. ANSI/IES RP-8 1983, *American National Standard Practice for Roadway Lighting,* IES. p. 3.

Chapter 4

Introduction to Basic Electrical Systems

Electrical/Hydraulic Analogy

The purpose of this book is to explain the function of electrical engineering as applied to legal proceedings. To convey this message effectively, it is helpful if the non-electrical engineer reader has a fundamental understanding of electricity as frequently encountered in our modern world.

The author has found that electrical terms such as voltage, current and resistance are difficult to understand for someone who is not schooled in these abstract concepts. Most people are comfortable, however, with ideas that associate with things that they are familiar with, and it is for this reason that the Electrical/Hydraulic Analogy is presented. A word of caution: there are pitfalls inherent in the use of analogies, and the reader is admonished not to depend strictly upon the comfort that this analogy offers.

The basis of the Electrical/Hydraulic Analogy is that electrical current is analogous to water flow. Electrical current is the flow of electrical charge, primarily electrons within a material. Water flow is the motion of water molecules within a pipe. Water flow is measured in units such as gallons per minute (GPM). Electrical current is measured in amperes (Amps).

The hydraulic force that produces water flow is pressure, and the analogous force that produces current is called electrical pressure. Water pressure is measured in pounds per square inch and electrical pressure is measured in volts.

As water flows through a pipe, it encounters roughness of the pipe surfaces and is restricted by the pipe cross section. This interference with water flow through the pipes or other conveyances causes an energy loss. Analogously, when electrons move through the atomic structure of a material, they encounter collisions and interferences from other atomic particles that produce a loss of energy. The resistance of an electrical material is measured in ohms. Resistance is used to account for the energy loss that occurs when current flows through a material.

The end product of water flow may be used for filling a tank, irrigating a lawn, or perhaps turning a water turbine for generating electricity. The practical end product of an electrical current flow may be for energizing a light bulb, creating heat, or powering a motor.

Table 4.1 Electrical/Hydraulic Analogy

Hydraulic System Element	Electrical System Element
Pump	Generator
Pressure (PSI)	Voltage (Volts)
Water Flow (GPM)	Current (Amps)
Friction of Water	Resistance of Wire (Ohms)
Pipe Size (I.D.)	Wire Size (AWG)
Valve (On/Off)	Switch (On/Off)
Water	Electrons

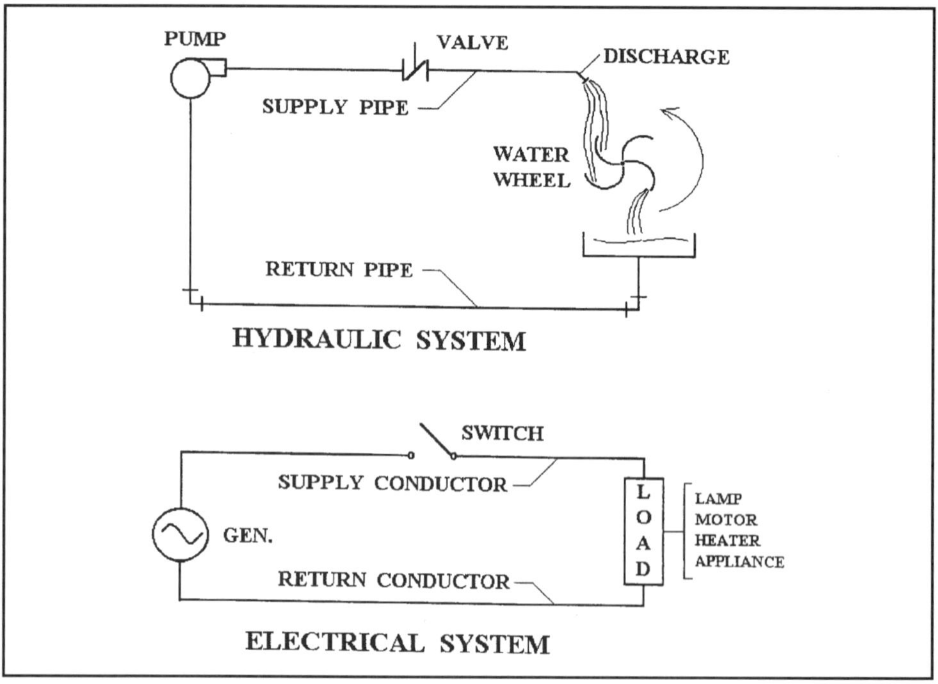

Fig. 4.1 *Electrical/Hydraulic Analogy.*

Ohm's Law

Perhaps the most satisfactory part of basic electrical science is that the real world behaves in a highly predictable manner if one is able to discover the applicable mathematical model and the parameters that apply to the case being considered. Electrical engineers with a modicum of humility are frequently amazed to find that Ohm's Law is universally upheld by nature. Ohm's Law may be stated in several equivalent forms:

$$V\ (volts) = I\ (amps) \times R\ (ohms) \qquad (4\text{-}1)$$

$$I\ (amps) = \frac{V\ (volts)}{R\ (ohms)} \qquad (4\text{-}2)$$

$$R\ (ohms) = \frac{V\ (volts)}{I\ (amps)} \qquad (4\text{-}3)$$

Basically, Ohm's Law states that when an electrical pressure of one volt is applied across a material whose resistance is one ohm, there will be one amp flowing through the material. Since there is current flow in a conducting material when a voltage is impressed across it, there will be energy lost as the current passes through the conductor.

The rate of energy loss in a material conducting current is called power and is expressed in watts. Power can be expressed in terms of voltage, current and resistance.

$$P\ (watts) = I^2 \times R \qquad (4\text{-}4)$$

$$P\ (watts) = \frac{V^2}{R} \qquad (4\text{-}5)$$

The power equation states that when one amp flows through a material having a resistance of one ohm, there will be one watt of power dissipated within the material. Another way of saying this is that when one volt is applied across a conductor having one ohm of resistance, there is one watt of power lost in the conducting process that ensues. One last relationship also exists for *purely resistive materials* such that power can be expressed as

$$P\ (watts) = V \times I \qquad (4\text{-}6)$$

Basic Electrical Circuits

For electrical energy to be usefully applied, there are three fundamental components necessary:

1. A source of electrical energy (*e.g.*, a generator, a battery, or other energy converter).
2. A transmitting means for carrying the energy from the source to the end use, such as wires or other conductors of energy.
3. An end use element, usually called a load (*e.g.*, a light bulb, an appliance, or a motor).

An electrical circuit incorporates each of these components. All basic electrical circuits can be treated as either series circuits, parallel circuits or a combination thereof.

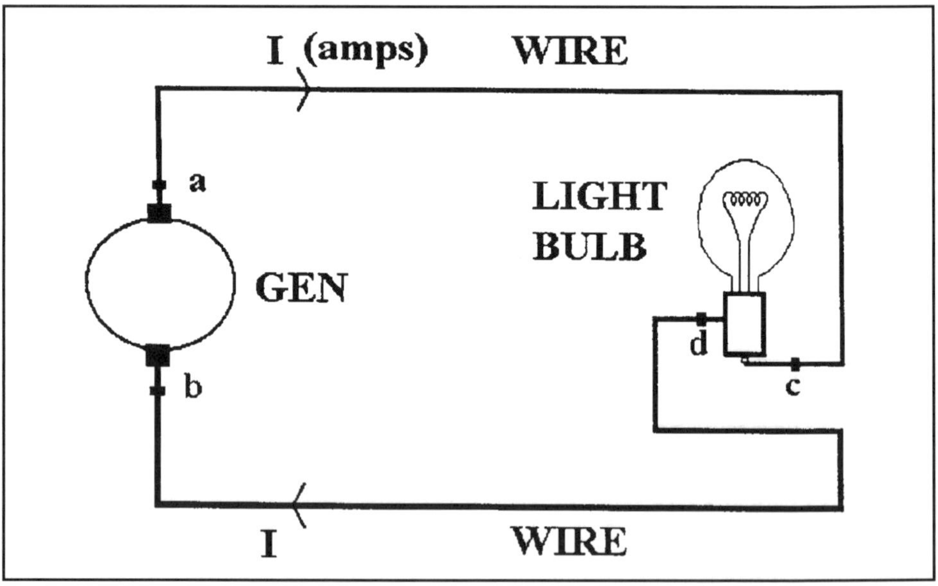

Fig. 4.2 *A simple series circuit.*

In the simple series circuit, the current flow I is the same through each component. As shown in Fig. 4.2, the current I (amps) which flows in the supply and return wires is the same current I which flows through the load and is the same current furnished by and returned to the generator.

Consider that otherwise there would be an accumulation of electrons somewhere which would eventually break through an electrical component similar to water breaking through a dam, creating a physical avalanche of current (a flood) not unlike a lightning strike. Since we know that such a phenomenon does not occur with ordinary current flow through a circuit in its normal mode of operation, we must reject this as a reasonable possibility.

It is the electrical pressure of the generator expressed in volts that causes the current I to flow through the wires and the load. Electrical engineers represent the elements of this simple series circuit schematically as shown in Fig. 4.3.

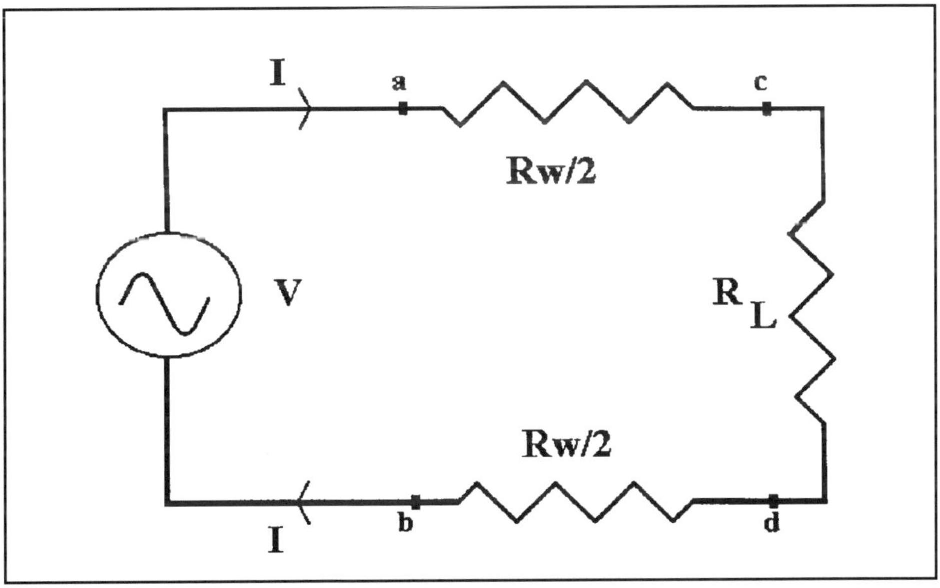

Fig. 4.3 *Schematic of a simple series circuit.*

In Fig. 4.3 the generator is schematically implied to supply an alternating current voltage (a.c.). The resistances of the supply and return conductors and the load are schematically represented by the resistor symbol shown on the right.

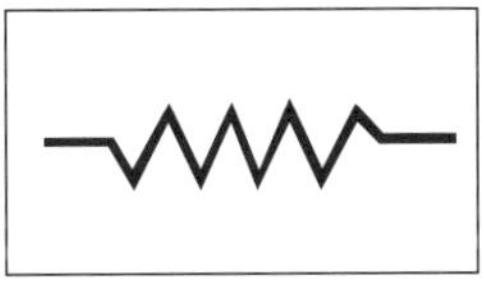

The values for the various electrical parameters of the circuit are shown algebraically by the symbols V, $R_W/2$, R_L, and I.

The total force or electrical pressure that produces a current flow through the wires and load must be divided among the component parts of the circuit. In accordance with Ohm's Law Eqn. (4-1), we may state that the supply pressure (voltage) is equal to the sum of the pressure losses (voltage drops) around the series loop.

$$V = I \times \frac{R_W}{2} + I \times R_L + I \times \frac{R_W}{2} \qquad (4\text{-}7)$$

Upon combining terms algebraically, we have

$$V = I \times (R_W + R_L) \qquad (4\text{-}8)$$

A practical example of the application of Eqn. (4-8) is a residential electrical circuit where a 100W light bulb is supplied by 100 ft. of copper wire (50 ft. each way from the source). In this example the electrical parameters are $V = 120$ volts, $R_W = 0.314$ ohms (14 AWG wire), and $R_L = 144$ ohms.

Solving Eqn. (4-8) for I gives:

$$I = \frac{V}{(R_W + R_L)} = \frac{120}{(0.314 + 144)} = 0.8315\ amps \qquad (4\text{-}9)$$

Using Ohm's Law Eqn.(4-4) to verify the power delivered to the load we find that

$$P_L = I^2 \times R_L = (0.8315)^2 \times 144 = 99.56\ Watts \qquad (4\text{-}10)$$

How can this be correct if the light bulb is rated at 100W? The answer lies in the fact that the light bulb is rated at 100W for a voltage of 120 volts. Since there is a small resistance in the wires of the circuit, there is some energy lost in heat as the current passes through these wires. There is then a voltage loss (pressure drop) between the source and the load. According to Ohm's Law, the voltage drop in the wires is given by

$$V_D = I \times R_W = (0.8315) \times (0.314) = 0.261\ volts \qquad (4\text{-}11)$$

The voltage at the supply V_{ab} = 120 volts is, therefore, reduced to a voltage at the load of

$$V_{cd} = V_{ab} - V_D = 120 - 0.261 = 119.74\ volts \qquad (4\text{-}12)$$

If this corrected voltage is applied at the load in Ohm's Law Eqn.(4-5), we find that

$$P_L = \frac{(V_{cd})^2}{R_L} = \frac{(119.74)^2}{144} = 99.57\,Watts \tag{4-13}$$

which agrees well with our previous calculation, Eqn (4-10). Mathematically Ohm's Law is upheld. Experimentally a circuit having these parameters will produce physically measurable results for current, voltage and power which agree closely with these calculations.

A simple parallel circuit is shown schematically in Fig. 4.4.

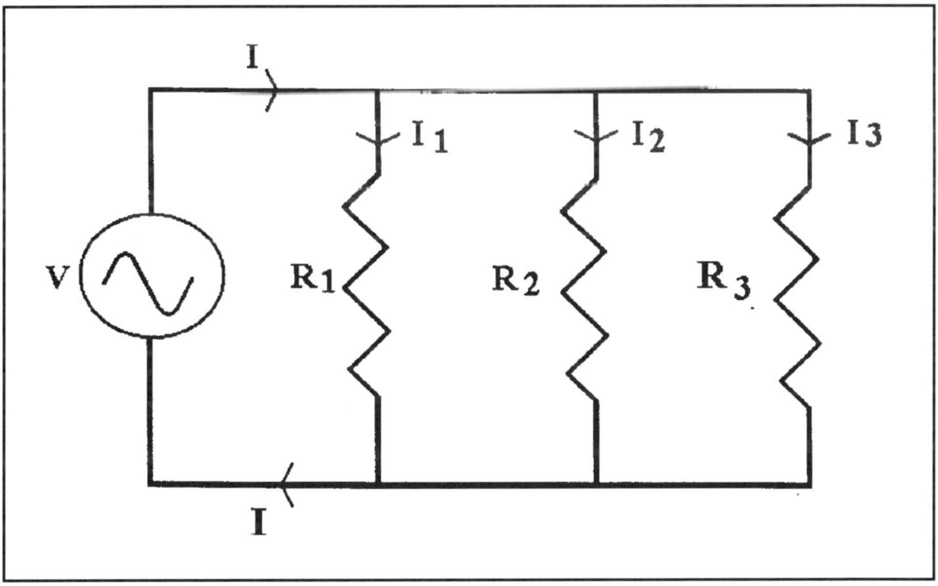

Fig. 4.4 *Schematic of a simple parallel circuit.*

In this schematic representation we have neglected the usually small resistances of the wires and only considered the relatively larger resistances of the loads. This representation is generally accurate when analyzing branch power circuits for supplying energy to residential and commercial users within a structure.

The principles of analysis for the parallel circuit are similar to those used for the series circuit and again we observe that electrons will not accumulate without limit in some element of the circuit.

Therefore, the current furnished by the generator must be returned to the generator.

Electrical engineers have been known to jest that electrical utilities have an ideal business: they loan the electricity to the customer and instantaneously get all of it (electrons) returned undamaged so that they can reloan the same electricity with interest. Actually the electrons furnished by the utility have greater energy (measured in kilowatt-hours) than do the returning electrons, which flow back to the utility's generators with zero net energy. The utility's generators then convert mechanical energy into electrical energy, thereby raising the energy of the electrons before they are resupplied to the customer. It is for this added energy value that the utilities charge.

A parallel circuit is most eloquently analyzed by recognizing that the supply current from the source must equal the sum of the parallel path currents furnished to the loads. Mathematically we have

$$I = I_1 + I_2 + I_3 \tag{4-14}$$

Applying Ohm's Law Eqn.(4-2) to Eqn.(4-14) gives

$$I = \frac{V}{R_1} + \frac{V}{R_2} + \frac{V}{R_3} \tag{4-15}$$

By algebraic combination we have

$$I = V \times \frac{(R_2 \times R_3) + (R_1 \times R_3) + (R_1 \times R_2)}{R_1 \times R_2 \times R_3} \tag{4-16}$$

While Eqn.(4-16) can be used to calculate the total current which is supplied to these three loads, the simplest method of analysis for parallel loads is to sum their individual current requirements. Since the voltage is assumed to be the same for each load, by Ohm's Law Eqn.(4-2) and using Eqn.(4-6) for resistive circuits, it is easiest to sum the power requirements of the individual loads and then divide by the supply voltage to determine the total current I.

Our electrical systems are by practice and by Code usually designed and constructed so that the supply voltage V remains essentially constant throughout the circuit. We are then relatively accurate in performing calculations based upon power requirements for loads rather than resistances. An example follows.

Calculate the line current required to supply ten 100W light bulbs connected in parallel at a rated voltage of 120 volts. The total load is first found:

$$P_L = 10 \times 100 = 1000\ Watts \tag{4-17}$$

Then by Eqn.(4-6) we obtain

$$I = \frac{P_L}{V} = \frac{1000}{120} = 8.333\ amps \tag{4-18}$$

Taking into consideration the wire resistance and placement of the loads, an accurate analysis for this example will be made. We can then compare the accurate analysis with the approximation obtained by Eqn.(4-18). Assume that the ten loads are located at 10, 20, 30, ..., 90, and 100 ft. from the source respectively. Then the total voltage drop to the furthest load can be shown to be*

$$V_D = 2.841\ volts \tag{4-19}$$

Since we assumed that the loads were equally spaced from the source, this total drop occurs in ten equal increments, and we can calculate the voltage at each load by the relationship

$$V_n = 120 - n \times \frac{2.841}{10} \tag{4-20}$$

where n = 1, 2, ...,10.

The current through each load can then be calculated, and the total current becomes

---

* The voltage drop in this example was calculated using a propietary program "Feeder Design" which was developed by the author's company *Nabours' Software*. For information about this software contact the publisher.

$$I = \sum_{1}^{10} \frac{120 - n \times 0.2841}{144} \qquad (4\text{-}21)$$

Evaluation of this summation for the ten load currents gives

$$I = 8.225\ amps \qquad (4\text{-}22)$$

The error between the more accurate value of *I*, Eqn.(4-22) and the approximate calculation for *I*, Eqn.(4-18) is

$$\Delta\ I = 8.225 - 8.333 = -0.108\ amps \qquad (4\text{-}23)$$

This is a 1.3% error, sometimes worth the more complex, time consuming, and accurate calculations.

Alternating Current Circuits

An a.c. electrical generator creates an electromotive force (voltage) from the interaction of magnetic fields as the rotor or rotating part of the generator mechanically passes the stator or stationary part of the generator. The power necessary to rotate the generator is converted by this magnetic field interaction from a mechanical source of energy, such as water from a dam or steam from a boiler, to an electrical source of energy.

Because the rotor windings or conductors of a generator pass cyclically in uniform rotational motion past the stator, the voltage produced by this motion can be described mathematically and confirmed by experiment to be a sinusoidal function of time as expressed by Eqn.(4-24).

$$v(t) = V_m \times \sin(\omega \times t) \qquad (4\text{-}24)$$

where ω is the equivalent angular rotation rate of the generator in radians per second, t is time in seconds, and V_m is the maximum voltage in Volts. In the United States ω is 377 radians per second, which engineers refer to as 60 Hertz (60 Hz) or 60 cps a.c. in older terminology. A plot of $v(t)$ for $V_m = 169.7$ volts, Eqn.(4-24) is shown in Fig. 4.5.

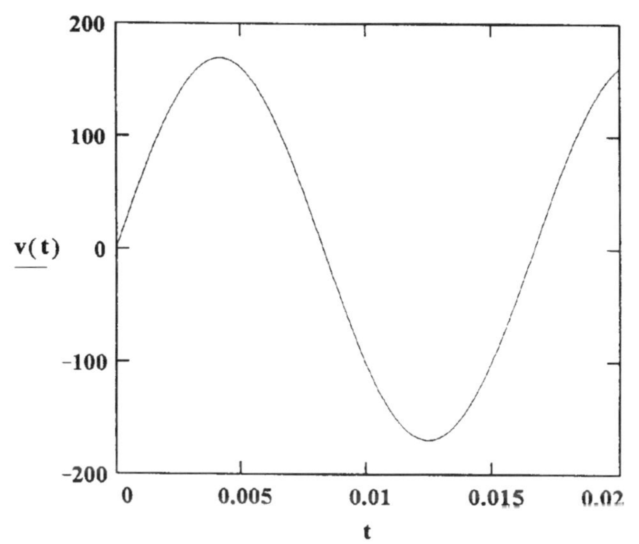

Fig. 4.5 *A.C. Voltage vs. Time. t := 0.0001 . . . 0.02; v(t) := 169.7 · sin (377 · t).*

The time required for one complete cycle of this voltage waveform is designated as T, where T is 1/60th of a second (0.0167 sec). There are 60 complete cycles per second from zero through positive values back through zero then through negative values and returning to zero.

This voltage variation, because it has both a positive and negative characteristic, causes the current which flows in an electrical circuit to also vary sinusoidally such that

$$i(t) = I_m \times \sin(\omega \times t) \tag{4-25}$$

The heating effect of an a.c. current flowing through resistance is governed by Ohm's Law Eqn. (4-1), however the voltage value that should be used in this equation is not immediately obvious. By using the Calculus we can show that the appropriate voltage to use for calculating the effective heating of an a.c. voltage is

$$V = \sqrt{\frac{1}{T} \times \int_0^T v^2(t) \times dt} \tag{4-26}$$

After substituting Eqn. (4-24) for *v(t)*, Eqn. (4-26) evaluates as

$$V = \frac{V_m}{\sqrt{2}} \cong 0.707 \times V_m \tag{4-27}$$

This is called the effective voltage or rms a.c. voltage. Rms stands for root-mean-squared, which is what the Calculus expression Eqn. (4-26) represents.

Similarly, we find that the effective a.c. current is given by

$$I = \frac{I_m}{\sqrt{2}} \cong 0.707 \times I_m \tag{4-28}$$

Effective values V and I are then used in the Ohm's Law equations for the mathematics of a.c. circuit analysis. As a matter of fact, we have already used the rms values of voltage and current in our previous examples of circuit analysis.

It is important to recognize that when we have a typical 120 volt a.c. electrical supply, the peak voltage for this source is given from Eqn. (4-27) by

$$V_m = \sqrt{2} \times V \tag{4-29}$$

or

$$V_m \cong 1.414 \times 120 = 169.7 \; volts \tag{4-30}$$

Generalized a.c. circuit analysis is more complex than the purely resistive circuit analysis which has been discussed previously. In generalized circuits it is necessary to consider two additional parameters when dealing with a.c. electricity.

The first element is termed inductance. Inductance is an electrical parameter which is applied like resistance, but which accounts for the storage of electrical energy in magnetic fields which are created whenever current is changing within a material. The inductive circuit element is represented by a resistive-like term called "inductive reactance" and is expressed mathematically by the following steady-state* equation:

$$V_L = X_L \times I \tag{4-31}$$

In Eqn. (4-31) V_L is the voltage across the inductive element (inductor), I is the current through the inductor and X_L is the inductive reactance. X_L is given by

$$X_L = j \times \omega \times L \tag{4-32}$$

where ω is again the sinusoidal radian angular frequency, L is the value of the inductance expressed in Henries and j is the imaginary operator $\sqrt{-1}$.

The second element is termed capacitance. Capacitance is an electrical parameter which is also used like resistance, but which accounts for the storage of electrical energy in electric fields which are created whenever a net electrical charge is stored temporarily within a material. The capacitive circuit element is represented by a resistance-like term called capacitive reactance and is expressed mathematically by the following steady-state equation:

$$V_C = X_C \times I \tag{4-33}$$

In Eqn. (4-33) V_C is the voltage across the capacitive element (capacitor), I is the current through the capacitor and X_C is the capacitive reactance. X_C is given by

$$X_C = \frac{1}{j \times \omega \times C} \tag{4-34}$$

where ω and j are as in the previous Eqn. (4-32) and C is the value of the capacitance expressed in Farads.

---

* The term steady-state is used by electrical engineers to refer to the circuit conditions that eventually obtain following the initial excitation of a circuit. For example, by closing a switch. In steady-state, a.c. currents and voltages behave sinusoidally as in Eqns. (4-24) and (4-25).

A simple series circuit having all three energy elements is shown in Fig. 4.6.

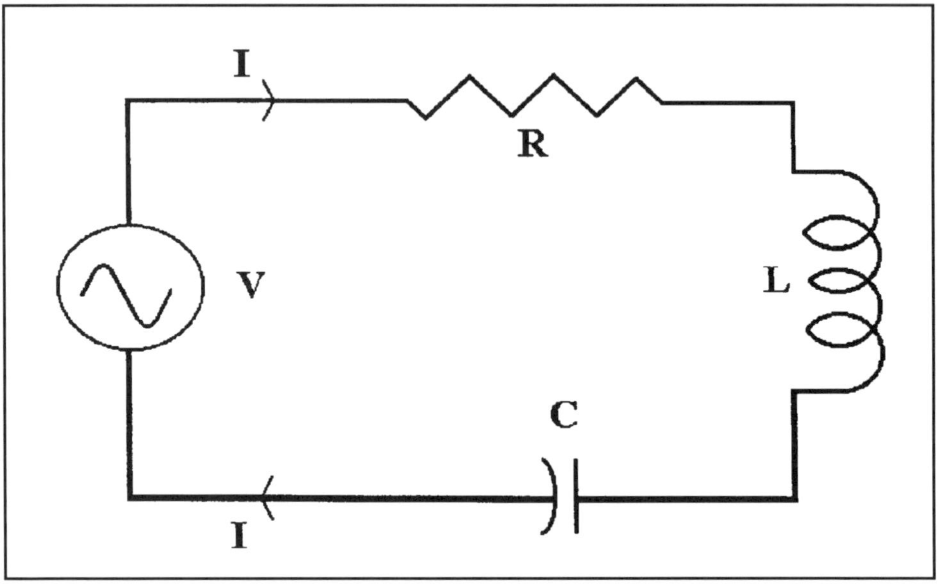

Fig. 4.6 *Series R-L-C Circuit.*

The analysis of this circuit requires the use of Complex numbers and Phasors, which are beyond the scope of this text. However, the results of such an analysis are important, and therefore, the following example is given.

Let V be a 120 volt 60 Hz source, let R = 1.5 Ω, L = 1.99 millihenries and C = 17,680 microfarads. These units have been chosen to be typical of values encountered in power circuits. The use of Ohm's Law, similarly to the derivation of Eqn. (4-7), gives

$$V = I \times R + I \times X_L + I \times X_C \qquad (4\text{-}35)$$

Finding X_L by Eqn. (4-32) and X_C by Eqn. (4-34) and substituting in Eqn. (4-35) we have

$$120 = I \times (1.5 + j \times 0.75 - j \times 0.15) = I \times (1.5 + j \textit{ times } 0.6) \qquad (4\text{-}36)$$

Solving for I we have

$$I = \frac{120}{1.5 + j \times 0.6} \qquad (4\text{-}37)$$

Using Complex number theory the current I may be expressed as

$$I = 74.28 \angle -21.8° \tag{4-38}$$

Eqn. (4-38) can be converted by using Phasor theory to a sinusoidal equation as follows:

$$i(t) = 74.28 \times \sin(\omega \times t - 0.38) \tag{4-39}$$

The -21.8° and its equivalent of -0.38 radians (rad) represent the fact that for this circuit, the current lags the voltage by this angular value, or by the time duration of 0.38/377 ≈ 1 millisecond. A plot of v(t) and i(t) for this circuit is shown in Fig. 4.7.

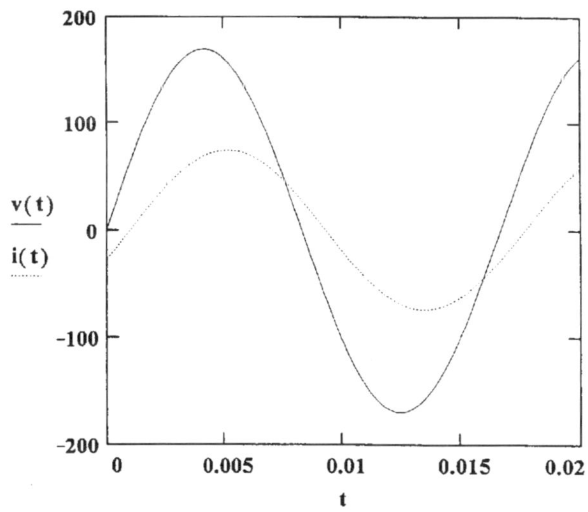

Fig. 4.7 *Voltage and Current vs. Time for R-L-C Series Circuit.*
t := 0.0001 . . . 0.02; v(t) := 169.7 · sin (377 · t); i(t) := (74.28) · sin (377 · t - 0.38).

An important effect of the reactive circuit elements, inductance and capacitance, on a.c. circuit behavior lies in the resulting time displacement of current with respect to voltage. *The use of Eqn. (4-6) is not valid for such circuits.* In fact, the correct version of Eqn. (4-6) for a.c. power circuits is

$$P\ (watts) = V \times I \times pf \tag{4-40}$$

where *pf* is the abbreviation for power factor. Power factor is defined as

$$pf = \cos(\theta) \tag{4-41}$$

where θ = the angle between v(t) and i(t).

For the example just considered, starting with Eqn. (4-35), the power factor is

$$pf = \cos(-21.8°) = 0.9285 \tag{4-42}$$

By Eqn. (4-40), the power dissipated in the series circuit we have analyzed should be

$$P = (120) \times (74.28) \times (0.9285) = 8{,}276.3\ Watts \tag{4-43}$$

This assertion can be checked using Ohm's Law which states that the power dissipated in the resistance of this circuit is given by

$$P = I^2 \times R = (74.28)^2 \times (1.5) = 8{,}276.3\ Watts \tag{4-44}$$

Depending upon the algebraic sign of the phase angle θ, the current in an a.c. circuit may either lag or lead the voltage. Physically it is impossible for | θ| to be greater than 90°; therefore, we have

$$0 \le pf \le 1.0 \tag{4-45}$$

Eqn. (4-45) states that the dissipated power in an a.c. circuit is always less than or equal to the apparent power for the circuit, where "apparent power" is defined as

$$P_{app} = V \times I \tag{4-46}$$

Common Electrical Power Systems

Residential electrical systems in the United States today are almost always furnished power by the local utility at 120/240V 1∅ 60 Hz a.c. The electrical utility supplies this voltage by transforming from a single phase (1∅) high voltage distribution circuit to the lower utilization voltage. The usual transformer arrangement for this process is called the Edison connection as shown in Fig. 4.8.

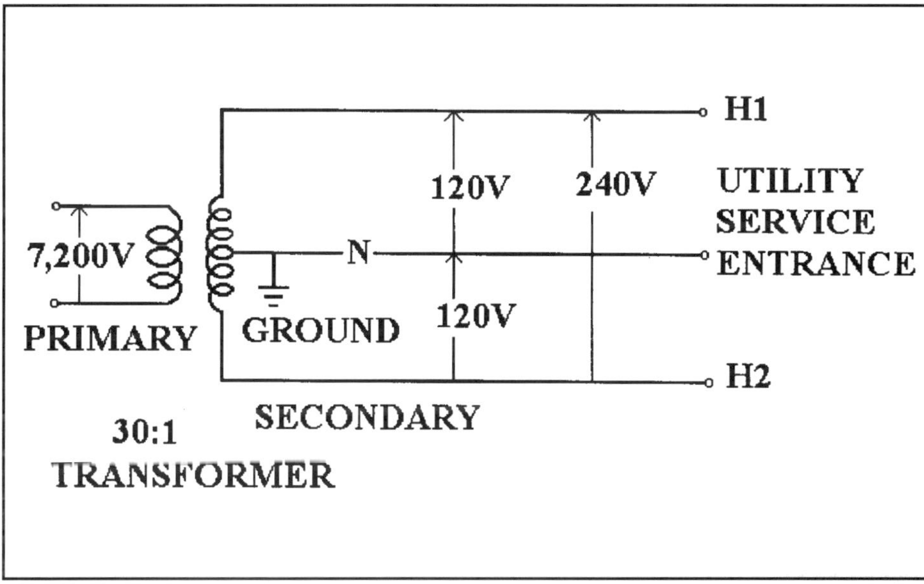

Fig. 4.8 *Residential Electrical Service—Edison connection.*

The utility may choose to use a lower or higher primary (PRI.) voltage, but the 120/240V secondary voltage has been standard since 1982.[1]

The Edison connection permits the user to supply loads of 120V by connecting the load from either of the "hot legs," H_1 or H_2, to neutral (N). By connecting between H_1 and H_2, a 240V load may be served. The grounded neutral at the center tap of the transformer's secondary enhances the safety of the system and establishes a stable reference for the secondary and primary electrical voltages.

If someone contacts either hot leg and a grounded conducting material, a 120V shock may result. The grounded neutral generally prevents a higher voltage contact and shock. However, if a person makes contact with both hot legs, say with the right and left hands, there will be 240 volts across his/her body. The effect of these contacts will be treated in detail later in this text.

The Edison connection has been referred to as a 115/230V 1∅ system or even in older terminology as 110/220V 1∅, but these systems are all the same configuration. The actual voltage supplied by the utility may vary, such that if the primary voltage is lower than 7,200V by say 10% (*i.e.*, 6,480V), then by the transformer voltage conversion ratio (30:1) the secondary voltage would be lower by 10% (*i.e.*, 108/216V). ANSI Standards have been established for the allowable voltage variation by U.S. utilities to assure the customer of a relatively stable supply voltage and, therefore, ensure the reliable operation of electrical products and appliances.

The most common commercial U.S. electrical system for the past forty or more years is the 120/208V Y 3∅ supply, called the 120/208V Wye system. This system is shown schematically for the three phase secondary transformer windings in Fig. 4.9.

The voltage relationships in a three phase power system are best explained using Phasor theory; however, one may think in terms of stationary vectors as shown in Fig. 4.10.

For the two vectors, *OA* and *OB*, shown in Fig. 4.10, the magnitude of the vector *AB* can be found using Trigonometry:

$$|\overline{AB}| = |\overline{OA}| \times \sqrt{\left(\frac{3}{2}\right)^2 + \left(\frac{\sqrt{3}}{2}\right)^2} = \sqrt{3} \times |\overline{OA}| \tag{4-47}$$

In the case where the magnitudes of *OA* and *OB* are both 120 volts, then the magnitude of *AB* = $\sqrt{3}$ x 120 volts = 207.8 volts. This is rounded to 208 volts for describing this power system as a 120/208V Y. The reader is encouraged to verify the preceding calculations and also invited to examine the other vector relationships between the three phase voltages in the Wye system.[2]

Industrial power systems generally utilize power at a higher voltage than do commercial users. The 277/480V Y system is commonly found in U.S. industrial plants. Note that 480 $\cong \sqrt{3}$ x 277.

In some cases of very large industrial power systems, even higher utilization voltages are found (*e.g.*, 4160Y/2400V and 12,470Y/7200V systems are used). Although these systems are Wye connected, they are conventionally designated in reverse numerical order compared to the lower voltage Wye systems. These voltage connections are also frequently utilized by electrical utilities for their overhead and underground distribution systems. The phase-to-phase voltage is again $\sqrt{3}$ times the phase-to-neutral voltage.

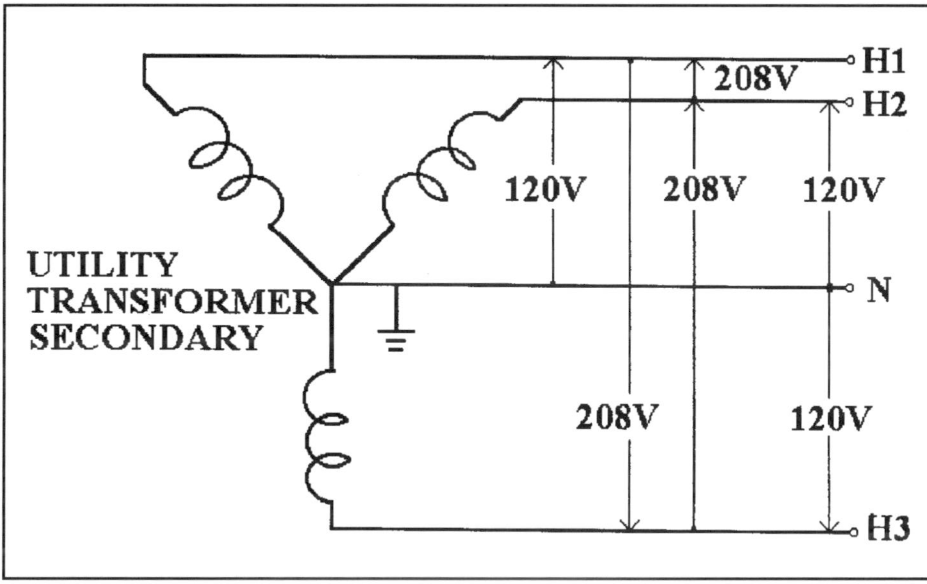

Fig. 4.9 120/208V Y Commercial Power System.

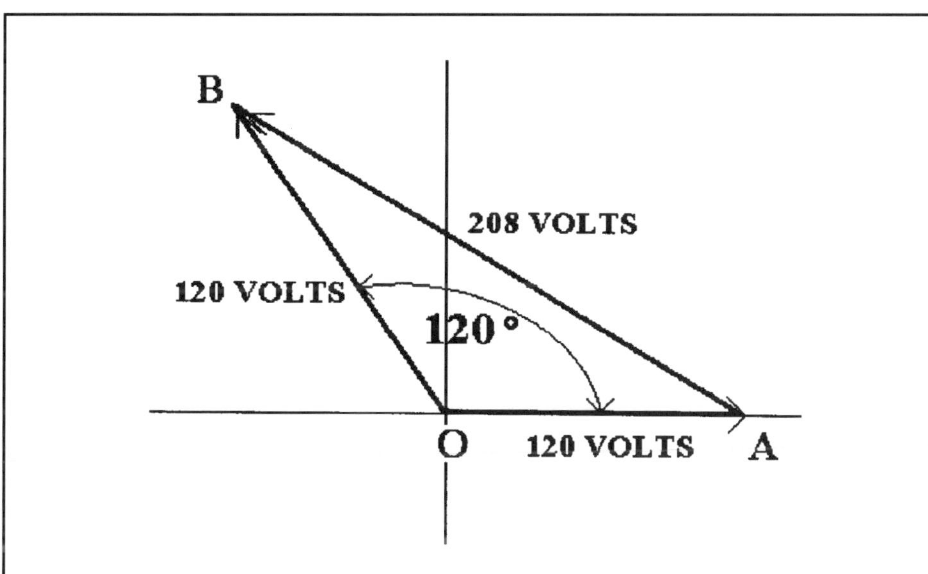

Fig. 4.10 Vectors for two phases in a three phase Wye System.

There are still other types of three phase voltage systems in use in the U.S. and around the world. The most common U.S. variation is the Delta system. Frequently commercial and small industrial users are supplied with a 120/240V Δ 3∅4W system as shown in Fig. 4.11.

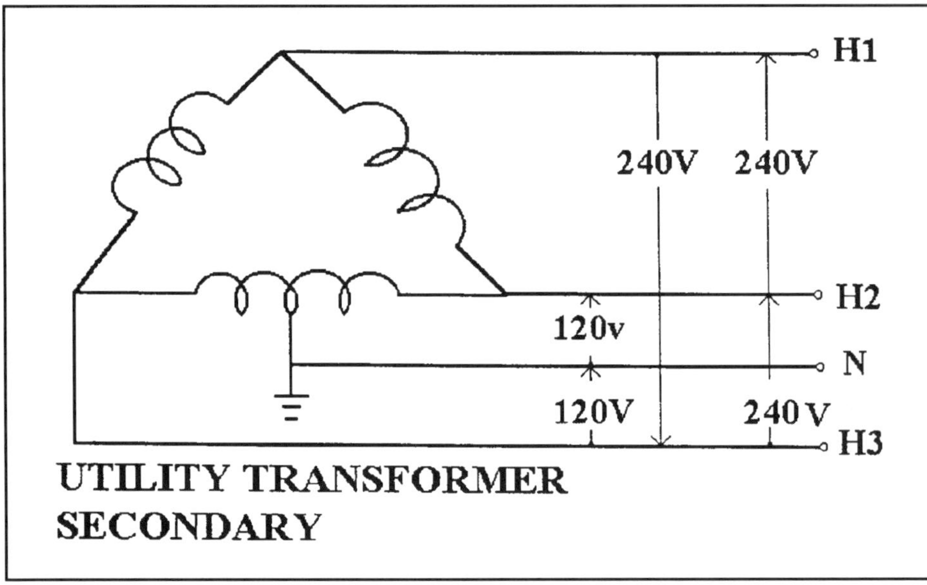

Fig. 4.11 *120/240V Δ 3∅ 4W Commercial Power System.*

In this 4W Delta system, the transformer winding with its center tap connected to neutral behaves like an Edison connection; however, because of the connections with the other windings this system can also be used to supply three phase 240V loads. There are many reasons for using three phase power systems, some of which are given below:

1. The energy delivered by a three phase system can be furnished with 1/3 less current per conductor than a single phase system, assuming the same voltage and load characteristics. Therefore, three phase systems can generally use smaller, less costly wiring than single phase systems. Even when the difference in number of conductors (3 vs 2) is considered, a three phase circuit will have 1/6 of the total line loss that a single phase circuit will have with the same size line conductors, voltage and load.[3]

2. The neutral current (fourth wire) in a three phase Wye system with balanced loads is zero. Therefore, under balanced or nearly balanced load conditions, there is zero or little energy loss due to the neutral or return current. Three phase motor loads require only three conductors for their supply in a Wye or Delta system.
3. Three phase motors use electrical energy at a constant rate; therefore, the torque production of a three phase motor is constant, not pulsating as is the case with single phase motors. Such motors run smoother with less bearing wear, are quieter and last longer.
4. Large horsepower motors are only produced in three phase rather than single phase because of their lower internal copper losses. Therefore, commercial and industrial users with requirements for large HP motors require three phase power.
5. Electrical utilities have established service requirements which specify load limits by class of service. Single phase 120/240V service is available for smaller loads, up to about 150KW. For larger loads, the utilities require three phase and specify what voltage will be furnished according to the load size. A large load user is therefore required to accept three phase service from the serving utility.[4]

Wiring Methods and Materials

Fixed wiring for branch circuit residential applications may consist of separately enclosed conductors in metallic or nonmetallic conduit (pipe), electrical metallic tubing (EMT), and/or surface metallic or nonmetallic raceway. By far, however, the most common method of fixed wiring for residences is nonmetallic sheathed cable, also known by the trade name of Romex. Commercial and industrial wiring systems are almost uniformly constructed using individual conductors enclosed in metallic raceways including rigid metal conduit, EMT wireway, cable tray, flexible metal conduit, and armored cable.

Conductors for the fixed wiring used in construction are manufactured in a wide variety of sizes (diameters) and insulation types. Wires are available in single and multiple conductor stranding, with the larger American Wire Gauge (AWG) sizes having only multiple stranding. Wire above #8 AWG, which is about 1/8 inch in diameter, is stranded so that the cables will be flexible enough to be pulled through conduit and will curve where the conduit bends. Construction wiring varies in size from #18 AWG (0.04 in. dia.) through 28 sizes up to 2000 kcmil conductor (1.63 in. dia.). The notation "kcmil" is read thousand circular mils.[5]

Ultimately, it is the rated temperature of a conductor's insulation that determines the maximum ampacity of the wire. The current carrying capacity of a conductor, called ampacity, is a function of the temperature developed by the $I^2 \times R$ loss of electrical energy into heat within the wire. For most applications, the National Electrical Code (NEC) has specified ampacities for conductors, and these values are generally accepted as national standards. The subject of Codes and Standards is treated in Chapter 3 of this text.

The operating temperature developed by a current carrying conductor is a function of several factors including the conductor material (*i.e.*, metal), the type and thickness of insulation, the type of installation (*e.g.*, within conduit, in free air, directly buried in earth, etc.), the ambient temperature and the mutual heating effect from other nearby conductors.

General wiring conductors for building construction are manufactured using either copper, aluminum, or copper-clad aluminum. They are available as bare wire and with insulation. The types of insulation available include plastics, mineral insulation material, rubbers, silicone-asbestos, plastic-asbestos and varnished fibers. The NEC recognizes insulation temperature ratings for general wiring of 60C, 75C, 85C, 90C, 125C, 150C, 200C and 250C. For most construction, fixed wiring insulation ratings of 60C, 75C and 90C are used.

Modern electrical wiring is usually moisture- and heat-resistant thermoplastic insulated PVC (polyvinyl chloride). This is the most popular choice for contemporary wire insulation because of cost and availability.

The distribution of electrical energy from the utility service to the various loads requires the separation of wiring into "branches" (branch circuits) which emanate from a distribution center. A distribution center contains circuit breakers or fuses to protect the branch circuit conductors from overloads or short circuits. A distribution center may consist of a panelboard, multiple fused switches, or a combination of such arrangements. A schematic for a three phase four wire circuit breaker panelboard is shown in Fig. 4.12.

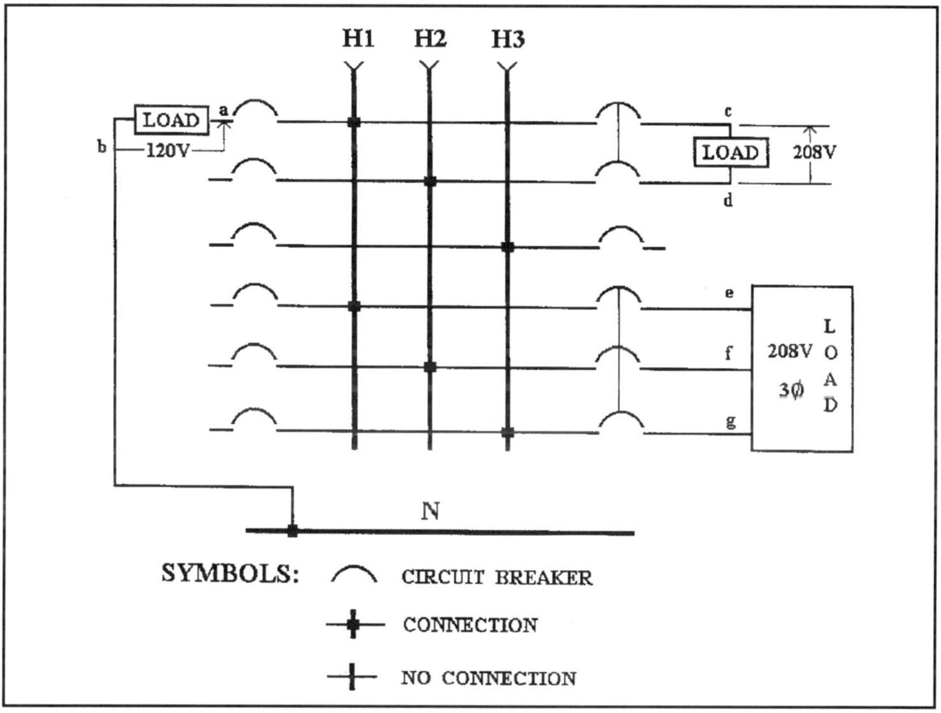

Fig. 4.12 *Schematic diagram for a 3∅ 4W Panelboard.*

In Fig. 4.12, H1, H2, and H3 are the three "hot legs" from a 120/208V 3∅ 4W service. These service voltages are available for use from the panelboard in several variations.

A 120V circuit can be created by connecting a branch circuit from any individual single pole breaker to the load and to the neutral as shown by connections a-b. A 208V 1∅ circuit will be formed when connections c-d are made from a two pole breaker to a 208V load. A 208V 3∅ circuit can be formed by making connections e-f-g from a three pole breaker to a three phase load. Whether or not a neutral is used, a separate grounding conductor is frequently installed along with the supply conductors to electrically ground equipment or appliances to the service ground.

Protection Devices

Overcurrent protection of the wires that emanate from a branch circuit distribution center can be realized with either fuses or circuit breakers. Fuses are thermal acting components whose fusible material is responsive to the heat produced by excess current, melting open and deenergizing the conductor upon overload. Fusible elements are made from Bismuth, Cadmium and Tin in various proportions to set the overcurrent limit and effect the opening of the circuit under extreme current conditions (*i.e.*, overloads or faults). Fuses are capable of interrupting specified current at a specified voltage rating. If either the voltage or current exceeds the specified ratings, a fuse may catastrophically fail under overcurrent conditions, rupturing or exploding. The maximum interrupting current specification for a fuse is termed the interrupting rating or interrupting capacity (IC). Typical IC's for fuses are 10kA, 100kA and 200kA, where kA = 1,000 Amps.

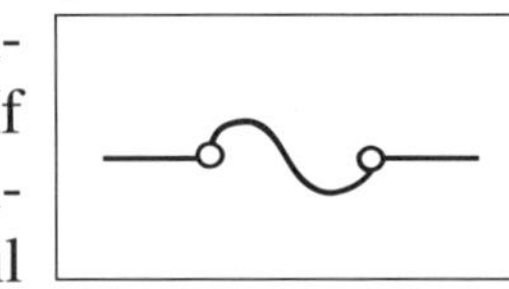

Older Edison base fuses were available in 15, 20, 25, 30, 35, 40 and 50 Amp ratings. These fuses no longer meet the NEC requirements but are available for replacement only.

Type S fuses are designed to make tampering or bridging difficult. They are installed with adapters to fit the Edison base socket. After installation the adapter cannot be removed without damaging the socket. Edison base fuses will not fit into the adapters and larger fuses cannot be installed in a lower rated adapter. Pennies will not fit into the adapters to by-pass the fuses.

Circuit breakers are automatically operated mechanical switching devices, usually of the thermal-magnetic type. The thermal element provides operation for lower current overloads, up to approximately 20 times the device rating. The magnetic element operates very fast (usually less than 20 milliseconds) for massive faults. Breakers are rated both for their normal current limit and for their ability to operate safely under high fault current. When the fault current exceeds the IC of the breaker, the breaker may not clear the circuit or it may fail catastrophically. Typical modern circuit breakers have interrupting capacities from 10kA to 200kA.

Circuit breakers are operable mechanically as switches and can be reset after automatic operation unless damaged by excessive fault current. Normal residential breakers are available in 10 to 225 Amp ratings with 18 ratings in between. 10kA is the most common interrupting capacity. Commercial and industrial circuit breakers are available in sizes from 100 to 4000 Amps.

Different manufacturers use various means to visually indicate that the breaker has automatically tripped, such as position of the toggle handle, or a red flag in a view window integral with the breaker housing. None of these indicators or their absence can be taken as positive proof that the breaker has automatically tripped.

Both fuses and circuit breakers have operating characteristics that are described by current vs time plots. Neither device will open the circuit instantaneously upon being subjected to an excess current. The time required for a fuse or circuit breaker to operate and open the circuit varies inversely with the current. This means that as the current increases in magnitude, the operating time for the protective device decreases. Table 4.2 gives typical operating times for generic fuses and circuit breakers vs current expressed in multiples of the device rating.

Table 4.2

Current Multiple of Rating	Fuse Operating Time (sec.)	Circuit Breaker Operating Time (sec.)
1X	∞	∞
2X	8	60
3X	2.2	20
5X	0.4	5
10X	<0.1	1
15X	—	<0.03

Common Appliances and Equipment

Electric heating equipment includes air heaters and water heaters, both fixed and portable. Portable residential air heaters are generally rated 120 V and less than 16 Amps (1800 Watts or less), and often include a fan for air circulation across the heating coils.

Commercial air heaters are available in much higher wattages and use larger fans for air circulation.

Air heaters may present a personnel or fire hazard if they are improperly located near combustibles or receive an inadequate air supply. UL listing indicates that suitable safeguards have been employed by the manufacturer against fire hazards resulting from appliance contact with draperies, furniture, carpeting, bedding and similar household combustibles. The operation of such a heater in an explosive or highly combustible atmosphere may be extremely hazardous.

Water heaters, including immersion heaters and high temperature kitchen sink units, are hazardous when operated without water or immersion in liquid. Heater element failure usually results in an open circuit, which is a fail-safe mode; however, a short to ground failure may produce a high resistance fault which can persist without activating the circuit protection. Such faults are likely to produce extended arcing or glowing at the fault point.

Ranges and built-in cooking units are similar to air heaters, except that the heating elements are usually encased in a steel sheath and frequently designed to operate by direct contact with a cooking utensil. Such heating elements can produce surface temperatures approaching 1200° F, and contact by personnel is dangerous. Naturally, contact with combustibles represents an extreme fire hazard. The author has observed an oven heating element, rated 3410 W at 240 V, fail by shorting to the grounded sheath. The element exhibited sustained arcing without tripping the 50 A circuit breaker. No incipient fire hazard existed, however, because the hot particles from the arc were confined to the interior of the oven. The situation could have been different if the faulted element had been a surface cooking unit.

Motor operated appliances in residences include refrigerators, dishwashers, garbage disposals, trash compactors, ventilating hoods and fans, washers, dryers, and room air conditioning units. Many of these motor-operated appliances also have heating coils or elements in conjunction.

The branch circuits to such appliances are subject to possible long term overloads because of frequent starting of the motors, which produces short term high current inrush, lugging of the motors or stalling, which also increases the line current draw, and increased line current caused by low voltage conditions.

A motor is basically a constant power device for constant mechanical load, so that input power is given by Eqn. (4-40)

$$P\ (watts) = V \times I \times pf \tag{4-48}$$

and the effect of reduced terminal voltage to a motor will be increased line current such that power remains constant. If the voltage becomes too low, the motor will stall or the line current will become too high and the circuit protection should operate.

Branch circuit wire size for motors is critical in order to avoid excessive heating of the conductors. The NEC, Art. 430 requires that single motor supply conductors be sized for the full load amperes of the motor on the circuit plus 25%. Certain exceptions are made for motors that will not operate continuously with load under any condition of use. In addition to this requirement, safe electrical design and construction practice is to select wire sizes that will avoid voltage drop in excess of 3% for the branch circuit and 5% from service entrance to the motor.[6]

The local utility power supply must also remain close to the nominal service voltage and should be within the ANSI standard supply range of 114 V to 126 V for a 120 V nominal supply under most operating conditions (*i.e.*, -5% to +5%, a band of 10% centered on 120 V). These standards and practices establish the minimum motor terminal voltage as 108 V for a 120 V system.

Most residential light fixtures are either incandescent or fluorescent. Both types may present a fire hazard due to excessive heat generation or arcing. An incandescent lamp (light bulb) operates by creating visible electromagnetic radiation (light) from a super heated tungsten filament that is encased in an evacuated or inert gas filled glass bulb. The incandescent lamp is a resistive load similar to a heater. Voltage variations, high or low, to an incandescent lamp cause current variations in approximately direct proportion according to Ohm's Law Eqn. (4-2).

$$I = \frac{V\ (volts)}{R\ (ohms)} \tag{4-49}$$

As voltage is reduced, lamp current, wattage and light output are reduced, and lamp life is increased. Conversely, if voltage is increased, lamp current, wattage and light output increase, and lamp life is reduced. These relationships are summarized in the following table given in percent of rated value:

Rated Voltage (%)	Rated Life (%)	Light Output (%)	Lamp Current (%)	Lamp Watts (%)
85	850	58	92	78
90	400	70	94	84
95	200	84	97	92
100	100	100	100	100
105	50	119	103	107
110	30	137	105	116
115	16	160	108	124

Incandescent light fixtures are rated for maximum lamp wattage. Although a standard base lamp socket used in a fixture may be rated 250 W, 250 V, the light fixture manufacturer may caution the user to limit the lamp wattage to a lesser value. This is done to avoid excessive heat transfer from the lamp to the fixture components, including wire and housing. High utility voltages can produce excessive temperatures within a light fixture, even though properly lamped, and create a fire hazard.

Fortunately, lamp life decreases more rapidly than lamp wattage and heat output increase so that sustained high utility voltage usually burns out incandescent lamps before fires are created. Note that incandescent lamps produce relatively high bulb surface temperatures even when operating at rated voltage and wattage.

Fluorescent lamps, on the other hand, are not high temperature light sources. The potential hazard for fire in fluorescent light fixtures resides in the ballast and wiring components of the fixture. Cold cathode (fluorescent) lamps require a ballast which accomplishes two functions: 1) it provides a high voltage, 400 to 1000 V transient (kick), to start the lamp by striking an arc, and 2) it limits the lamp current to a safe level that will not cause the gas in the fluorescent tube to burn itself out immediately after the arc is struck.

A fluorescent ballast circuit usually involves a transformer, a capacitor and perhaps an inductor with the associated wiring. These components are encased in a metal can, the ballast housing, which is filled with an asphaltic-sand potting compound. The potting compound serves as an insulator for both heat and electricity as well as a sound deadening material. The heat produced within the ballast, which can cause the case to approach 90° C (194° F), must be transferred to the surrounding environment in order for the wiring and components within the fixture and ballast to operate safely. Since 1978, the NEC has required all except simple reactor ballasts to have thermal protection integral within the ballast. Such fixtures are termed Class P and are required to be so labeled. Before Class P fluorescents were being produced, a large number of fires were traced to malfunctioning and overheated fluorescent ballasts. Since 1978, the incidence of fires attributed to fluorescent fixtures has fallen dramatically.

Radios, television sets, stereo systems and computers have all become frequently used appliances in our residences. These appliances do not behave electrically like simple resistive loads, nor do they have the characteristics of motors. All such appliances produce operating heat and must be properly ventilated to avoid fire hazards.

Television sets and computers, in addition to heat, also operate with cathode ray tubes which require high voltage electronic circuits. High voltage, in excess of 600 V, can stress insulation, causing wiring to fail internally in the appliance and can produce arcing. Usually the protective devices within such appliances deenergize failing circuits and avoid fires; however, misapplications, dirty conditions and the proximity to combustibles can lead to fire.

The high technology engineering and relative cost of electronic appliances when compared to hair dryers, clothing irons and toasters, for example, tends to preclude the catastrophic failure of such products. Nevertheless, electronic appliances, even the highest quality, can and have caused fires.

Products such as toasters, toaster-ovens, coffee makers, hair dryers, hair curlers, heating pads and electric blankets all belong to the same classification of small heating appliances. The previous discussion of heating equipment has implications regarding these appliances. The elevated temperature surfaces of such products must obviously be guarded from personal contact or close proximity to combustibles.

The electrical shock hazard of hand held equipment is pronounced, and since 1971 the NEC has recognized this electrocution hazard by requiring ground fault interrupting (GFI) circuit breakers and receptacles for certain work place and residential locations where water is likely to be present.

GFI protection, which for personnel operates at 4 to 6 milliamps (mA), is sensitive to the flow of current between an energized conductor and ground no matter what that path may be including ground wire, metallic conduit, metallic water pipe, or metallic building material. Beginning on January 1, 1991, for hand held hair dryers, UL 859 has required the use of 1) an integral ground-fault circuit interrupter (GFCI), or 2) an integral immersion detection circuit interrupter (IDCI), or 3) an integral protection device of another type that de-energizes all current-carrying parts when the appliance is immersed in water having an electrically conductive path to ground. U.L. has termed all such integral ground-fault interrupters for appliances Appliance Leakage Current Interrupters (ALCI) and requires this protection to activate when the appliance leakage current to ground is in the range of 4 to 6 mA.

The GFI device is not sensitive to phase-to-phase or phase-to-neutral faults (*i.e.*, from one hot leg H1 to the other H2 or from one hot leg H1 to neutral N). When such faults occur on a GFI protected circuit, the shock and fire hazards are not reduced by the GFI device. Only the basic level of circuit breaker or fuse protection is effective under phase-to-phase or phase-to-neutral faulting. For most residential receptacle, lighting, and appliance branch circuits, the probability of phase-to-phase faulting is low. There remains, however, a potential for such faults to occur even on a GFI protected circuit.

Manufacturers of ground fault circuit interrupters recommend that these devices be checked on a regular basis, as often as once a month. The GFI is equipped with a TEST button that connects a resistor between hot leg and ground to produce a ground fault current of not more than 6 mA which should immediately trip the breaker to deenergize the down stream circuit. A RESET button, which pops out under test, is used with the GFI to restore power to the circuit following the test. Some manufacturers recommend testing the down stream devices or loads to be sure that the power has been interrupted even after the RESET button pops out.

Chapter Notes

1. ANSI/IEEE Std. 141-1986, *IEEE Red Book,* 6th Ed., The Institute of Electrical and Electronics Engineers, Inc.

2. Brenner and Javid, *Analysis of Electric Circuits,* McGraw-Hill, 1959. Ch. 17.

3. Ernst A. Guillemin, *Introductory Circuit Theory,* John Wiley, 1953. pp. 388-392.

4. Tucson Electric Power Co., *Electric Service Requirements,* Tucson Electric Power Co., Tucson, AZ.

5. ANSI/NFPA 70, *National Electrical Code-1993,* NFPA, Batterymarch Park, Quincy, MA. Art. 310.

6. Ibid. Art. 210-19 (a) (FPN), 310-15 (b) (FPN), and 215-2 (FPN).

Chapter 5

Electrical Energy Effects on Humans

Introduction

Electrical energy exists on earth today due to both natural and man-made phenomenon. The presence of natural electrical energy was undoubtedly first recognized by the manifestation of lightning. The devastating effects of nature's power as evidenced by lightning must have been awesome to the earth's first cognizant life forms and was likely identified as a product of the gods who ruled the earth's environs and dictated man's existence on this planet.

The first practical relevancy of electrical energy to man was probably lightning-produced fire which prehistoric humans adapted for useful purposes. Throughout the history of man's evolution and increasing recognition of electrical phenomenon, many observations and curious inquiries must have arisen. Imagine the first human's discovery that an inanimate object, human hair, would move untouched toward a bone comb or that two pieces of rock would rotate untouched in mysterious ways when brought close to one another. These discoveries and inquiries are the essence of a mystery that even today continues to challenge mankind's ability to explain the phenomenon of electricity.

Static electricity was first studied scientifically by the Greeks who were perceptive in their observations that electricity had at least two forms, positive and negative, and that these forms tended to attract or repel one another according to their polarity.

Magnetism was also recognized by this time and it was known that certain natural materials not only possessed the properties of expulsion and attraction between themselves but exhibited, as well, a tendency to align their axes in relation to the earth so as to always exhibit a geometrically directional preference.

The work of Coulomb in 1785 first quantified the electrostatic force that existed between charged bodies. In the early 19th century,

Oersted observed that a magnetic field exists in the space that surrounds a wire carrying an electrical current, and Andre Ampere found that magnetic forces exist between two current carrying wires. In 1831 Faraday announced his discovery of electro- magnetic induction[1] and formulated a mathematical law for the voltage induced in a conductor due to a time varying magnetic field. Faraday's discoveries formed the theoretical basis for electric transformer operation and for modern mathematical analysis of induction currents. It was not, however, until the later part of the nineteenth century that scientists discovered electromagnetic waves and that the propagation of such waves was proposed as a model for the transference of electrical energy through space.

Electrical energy, both natural and man-made, has and does affect life on the planet earth. Electricity has become all pervasive in the modern human lifestyle. Principally humankind has found electricity to be an extremely useful and life enriching form of energy, readily available, economically feasible at least in developed countries, and highly versatile.

There are negative aspects, however, to the widespread applications of man-made electricity. On the negative side, electrical energy can be extremely dangerous whether in the form of lightning, high voltage power lines, household wiring and electrical appliances, industrial tools and processes, or electromagnetic radiation. Three known mechanisms by which electrical energy can interact with the human body are:

1. Direct contact current or close proximity arcing current flow from an energized source of electricity through the body.
2. Directly induced current flow within the body caused by close proximity to time varying electrical energy sources.
3. Electromagnetic waves both ionizing and non-ionizing that produce biological effects in humans.

Each of these mechanisms will be treated in detail in the following sections of this chapter.

Direct Contact Effects of Electricity on Humans

During the bodily contact with an electrical energy source of the type found throughout the U.S. in homes, offices, and commercial and industrial buildings, it is the passage of 60 Hertz electrical current through tissue and organs which can produce adverse physiological effects upon the body.

The flow of current can be related to the voltage across the body by Ohm's Law:

$$I = \frac{V}{R} \tag{5-1}$$

where I is expressed in amperes, V in volts, and R in ohms. Resistance of the body (R) is a function of several factors including:[2]

1. Skin condition—abraded or broken skin offers a lower resistance path between the epidermal skin and the vascular corium and deep tissues. Perspiration, water, saliva and blood all can reduce the surface resistance of skin contact. Calloused skin and the presence of keratin in the skin increases resistance.
2. Contact resistance—firm contact pressure between the skin and an energized conductor can reduce resistance. Conversely, light pressure increases contact resistance.
3. Area of contact—the resistance of skin contact is lowered as the contact area is increased and the heating effect of current flow varies inversely with the area of contact between energized conductor and the skin.

Typical values of resistance used to estimate electrical current flow through the human body vary from 500 ohms to 2000 ohms when the various factors stated above are taken into consideration. Journeymen electricians have frequently been known to lick their fingers before touching a 120 volt conductor to test it for an energized condition. Evidently their hands are heavily calloused and their "startle reaction" is much more insensitive than the average person's.

To place this in numerical perspective, the following current flows are calculated for a 120 volt rms contact between both hands, one to an energized conductor and the other to a grounded surface:

R (ohms)	I (milliamps)
500	240
1000	120
1500	80
2000	60

The first recorded death from commercial electricity was in 1879, when a stage carpenter in Lyon, France, was killed by electrocution from a 250 volt Siemens dynamo.[3] The first electrical execution took place at the state prison in Auburn, New York, on August 6, 1890. A convicted murderer named William Kemmler, who had taken an axe to his lover Matilda "Tillie" Ziegler, was sentenced to die in one of the first three electric chairs constructed for that purpose.

Inexperience at the time led to the condemned man receiving a fifteen second shock at an unrecorded current magnitude. It was greatly disturbing to the medical doctors, prison officials, and witnesses to this execution that Mr. Kemmler did not immediately die after this attempted electrocution. A second shock was administered for a "considerably longer time," and the prisoner finally died. An account of the witnesses' reactions to these proceedings reports that "The witnesses may have nearly wanted to die too, after this cruel surprise. The electrocution, complete with heaving chest, gurgles, foaming mouth, bloody sweat, burning hair and skin, and smell of feces, made a stunning impact on them. Many who had pulled political strings to secure a witness seat regretted doing so."* Despite the public condemnations of the process that quickly followed this first execution by electrical current, even the severest critics agreed that it appeared painless, the discharge apparently rendering the victim unconscious well before he could feel pain.

---

* James F. Penrose, "Inventing Electrocution," *American Heritage of Invention & Technology,* American Heritage, a division of Forbes, Inc., Spring 1994, Vol. 9, No. 4. pp. 35 to 44.

The effect of current upon the human body has since then been studied scientifically by Dalziel.[4] The factors influencing ventricular fibrillation, which is the most common form of death arising from electrical contact, are body weight, current magnitude, and shock duration. Dalziel found empirically that the following equation can predict the ventricular fibrillation (v-f) current for man:

$$I = \frac{k}{\sqrt{t}} \tag{5-2}$$

where I is expressed in milli-amps, t in seconds, and k is an empirical constant varying with body weight from about 100 to 185 lbs. The lower values of k apply to children and a 155 lb. man would have a k value of approximately 165. The valid range of time for Eqn. (5-2) is from 8.3 milli-seconds to about 5 seconds. This calculated current will produce v-f in about 1 out of 200 individuals. Tests have also indicated that this threshold v-f current equation holds for t as long as 20 or 30 seconds.[5] The threshold v-f current predicted by Eqn. (5-2) is valid for intact individuals who make ordinary touching contact with energized conductors. An individual in a hospital with a catheter or instrument probe penetrating the skin will have a much lower v-f current.

Dalziel[6,7] has concluded from studying accidents involving impulse shocks (*i.e.,* current pulses of brief duration), that an electrical impulse having an upper limit of 50 Joules represents the safe level for such shocks to humans. The duration of the impulse in these studies was approximately 0.2 sec. with an off period of approximately 0.8 sec.

The Joule is a metric measure of energy equal to one newton-meter which is also equal to one watt-second. When the voltage and current are in-phase, one watt equals one volt-ampere. For the durations of impulse shock described above, the energy delivered by a 60 Hz a.c. source is essentially given by volts times amps. Therefore, for a 120 volt rms source and a 500 ohm load, the shock duration of 0.2 sec. represents 5.8 Joules of energy and should be well below the limit of safe exposure. As a practical matter, UL has established standards for electric fence controllers which follow these guidelines.

If the effects of increasing current through the body are examined, we find that the following ranges prevail:

9.5 to 22 milli-amps	let-go limit for 98% of adult men
20 to 40 milli-amps	extreme muscle contractions, breathing difficulty
50 to 250 ma	respiratory muscle spasms can lead to ventricular fibrillation within 3 to 5 seconds
250 ma to 4 amps	heart stops. If exposure is only a small fraction of a second, the heart may resume beating spontaneously. If exposure is long enough[8] (7 sec. at 2300 volts, followed by 52 seconds at 550 volts in a legal electrocution) the damage is permanent

The let-go limit represents the maximum current flowing through the hands or hand of an individual for which the involuntary contraction of the muscles of the hand can still be overcome by the voluntary commands from the brain. For current above the let-go value, the brain can no longer control the contraction of the hand and an individual cannot turn loose from a conductor or wire. For continuous contact with 60 Hz a.c. electricity, it has also been determined that ventricular fibrillation occurs at a threshold of approximately 10 to 50 Joules with current in the range from 20 to 250 ma.

Physical evidence of electrical contact can include skin trauma such as burns. The estimated lowest current that can produce noticeable first or second degree burns in a small area of the skin is 100 ma for 1 second. Burns caused by the direct heating effect of electrical current (I^2R) frequently conform in size and shape to the metallic conductor of contact. In a study of 220 electrocutions of which 50% were caused by contact at less than 1000 volts, 45% of these had no perceptible electrical burns. In contrast, of the 50% of the contacts at higher than 1000 volts, only 4% had no perceptible burns.[9]

Directly Induced Current Flow Within the Human Body

The presence of a time varying electromagnetic field (EMF) in proximity to a material that conducts electricity will result in free charge displacement due to the electric field (E field) forces on such charges and conduction currents due to the magnetic field (H field) forces on the moving charges within the material. The human body has both free and moving electrical charges that exist within the cellular structure of the body tissue and organs. Therefore, when an external EMF of sufficient strength is present in the human environment, there will be displacement and conduction charge motion induced within the body.

The calculations of electrical induction on inanimate objects is a precise science utilizing well established mathematical models and subject to experimental verification. Such calculations for induction in the human body are not however easily made nor subject to simple experimental verification. Mathematical models for the body in proximity to an EMF source are subject to a great many simplifying assumptions and idealizations. Nevertheless, such calculations can be performed and they yield informative and meaningful results which provide engineers and biomedical scientists with insight as to the physical effects of such induction on the human body.

All energized electrical power lines have electric and magnetic fields associated with their supply of electrical energy. An electrical power line is energized by a voltage source at the generating (sending) end of the line and connected through conductors to the load at the other (receiving) end of the line. In order for energy to be transferred from the generator to the load, an electric and a magnetic field must be formed in relationship to each other and as physically determined by the power line conductors including their geometry.

The line's electric field is produced by the generator and the magnetic field is formed when the load is connected. The current which flows in the line conductors is a result of the energy transfer occurring due to the electric and magnetic field interactions, but the current flow in the conductors is not transferring power to the load. Rather, the line current causes a loss ($I^2 \times R$) in the conductors. In order to minimize the line losses in a power line the current must be held to reasonable values and the resistance of the line must be economically minimized.

The significant fact about the power line EMF phenomenon is that the energy flows through the air surrounding the line, truly creating an electrical power corridor in space. The line conductors serve as an energy guide for the directional transfer of power from source to load. Anything which resides within this EMF corridor will be subject to the effects of the energy transfer taking place in the space around the power lines. If the material residing within the corridor has an electrical nature, electrical charge within the body of the material may be displaced or realigned. It is ultimately the effect of such electrical activity that scientists are searching for as biological effects of power line EMFs.

The EMFs of overhead power lines, 5 kV to 1000 kV, vary with distance from the conductors and with the shielding effects of structures and natural landscape features such as trees and hills. Some examples of the maximum electric fields at ground level found near power lines are as follows:

Line Voltage	Max. Field at Ground Level
23 kv	50 V/m
115 kv	2 kV/m
345 kv	5.6 kV/m
500 kv	8 kV/m

These electric fields are reduced by increasing distance from the lines. For example, a 115 kV transmission line (TL) with 2 kV/m maximum field, will produce a field of approximately 200 V/m at 65m (213 ft) from the center line of the TL corridor.

The magnetic fields associated with a power line can be related to the line currents flowing within the conductors and, as is the case with electric fields, these magnetic fields will also decrease with increasing distance from the line. An example TL with line currents of 2000 Amps and line conductors at 18.5 m above the ground will produce a maximum magnetic field of 0.18 Gauss* (G) at ground level directly under the line. This field will decrease to approximately 0.02 G at a distance of 80 m (262 ft.) from the center of the line.

* The Gauss is a unit of magnetic flux density which measures the ability of electric currents to produce magnetic induction forces at a point.

The Electric Power Research Institute (EPRI) has set forth a set of six characterizing parameters which are of importance when considering the electric-field induction on objects in proximity to high-voltage energized power line conductors:[10]

1. The short-circuit current I_{sc}, which is the current that is induced in a zero impedance path from object to ground.
2. The voltage V_{og}, which is the voltage that is induced between object and ground and depends upon the impedance between the object and ground.
3. The impedance between object and ground given by

$$Z_{og} = V_{og}/I_{sc}$$

4. The capacitance between the object and ground, C_{og}.
5. The spark-discharge capacitance to ground, C_s, which has a negligible series resistance in the spark-discharge time of 1 microsecond.
6. The maximum energy stored per half-cycle in the object

$$J = \frac{1}{2} \times C_s \times (\sqrt{2} \times V_{og})^2 = C_s \times (V_{og})^2 \qquad (5\text{-}3)$$

The EPRI has also published tables for use in applying these parameters to various objects including people, horses, vehicles, houses and other structures in uniform electric fields.

An important effect of induced voltage on an inanimate object in the vicinity of an electromagnetic field is the contact by humans standing on the ground with the object after it becomes energized. If the V_{og} of the object is high and the Z_{og} of the object is high with respect to the impedance between human contact and ground, then the current flow through the person can be hazardous.

Most objects that are in close proximity (within 3 meters) to high voltage power lines are low impedance structures such as metal towers or wooden poles so that Z_{og} is low and even if a person standing on the ground touches such a structure, the current flow through the human is insignificant (*see* parameter 3 above and p. 78).

The danger from high voltage overhead power lines to people standing under the lines due to induced voltage or current is minimal so long as a third factor such as a very large inanimate object in the vicinity of the line does not enter into the situation. If an ungrounded large object is affected by induction from a high voltage power line and a human makes contact with the object while standing on the ground, then it is possible for an electrical current to flow through the person with an adverse effect.

Formulas for electric field induced short-circuit currents, I_{sc}, are given in the EPRI Transmission Line Reference Book,[11] Tables 8.8.2 and 8.8.3, for a number of geometric shapes and specifically the following:

Cow	$I_{sc} = 17.5 \times 10^{-9} \times h^2 \times E$
	$C_s = 200\ pF$
People	$I_{sc} = 5.4 \times 10^{-9} \times h^2 \times E$
	$C_s = 140\ pF$
Large School Bus	$I_{sc} = 3.9 \times 10^{-7} \times E$
	$C_s = 1800\ pF$

For each of these objects, in a 7 kV/m electric field, the following short-circuit currents can be calculated:

Cow	$I_{sc} = 0.17\ mA$
Person	$I_{sc} = 0.13\ mA$
Large School Bus	$I_{sc} = 2.73\ mA$

In the case of the large school bus, when stopped on dry pavement under and perpendicular to a 500 kV transmission line, if a person standing on the ground with wet shoes or on wet grass were to make hand contact with the bus, there would be a current passing through the body of 2.73 mA. This is a large enough current to startle 20% of adult men and in fact exceeds the perceptible current threshold for almost all humans. Obviously, in this example the electric field induced effect on humans is undesirable but not life threatening unless the startle reaction creates an inadvertent danger such as stepping backward into traffic or tripping.

A greater danger from the startle reaction caused by electric field induced voltages on persons lies in the spark discharge current which occurs slightly before full contact is made by a person who reaches out to touch a grounded object.

The voltage from person-to-ground V_{oc}, is determined in a spark discharge situation by the electric field, the individual's position and the impedance to ground. In the case of very high impedance to ground, the voltage to ground is

$$V_{oc} = \frac{I_{sc}}{\omega C} \tag{5-4}$$

where C is derived from EPRI measured data, Fig. 8.10.14,[12] and I_{sc} is as given for people on p.15 preceding,

$$I_{sc} = 5.4 \times 10^{-9} \times h^2 \times E \tag{5-5}$$

where h = height of person in meters and E = electric field intensity at ground level in V/m.

For the I_{sc} calculated in the previous example with a 7kV/m electric field and h = 1.855 m (h = 6 ft) we have a typical V_{oc} of

$$V_{oc} = \frac{0.13 \times 10^{-3}}{377 \times 160 \times 10^{-12}} = 2{,}155\ V\ rms \tag{5-6}$$

This is a peak voltage at discharge of $2{,}155 \times \sqrt{2} = 3{,}048\ V$. To place this high voltage in perspective, the dc voltages attained by people walking on carpet in a very dry environment typically range from 4 kV up to 8 kV. Everyone has probably experienced the startle effect that results from the discharge of the accumulated static electric charges at these dc voltages.

A significant difference in the ac induced spark discharge situation lies in the possibly repetitive nature of the discharges for slowly made or brushing contacts. Such spark discharges are undoubtedly more startling than experienced in a static carpet induced single discharge.

Magnetic fields at power line frequencies may cause significant voltages to be induced in objects whose length is considerable and oriented parallel to the power line. Such objects as metal fences, pipes and wires are known to exhibit large induced voltages when placed parallel and in close proximity to overhead power lines. Many years ago farmers discovered that by coiling a number of turns of wire under an overhead transmission line crossing their farms that they could steal electrical power and electrify their barns. Unfortunately some of these attempts to obtain free electrical power burnt the barns down or electrocuted someone or some animal.

We cannot readily calculate the induced voltage in a conductor from a measurement of the magnetic field strength, or magnetic flux density $B = \mu H$ in Gauss. Rather, it is the magnetic potential, M, which is analogous to space potential that is used to calculate electric field induced voltages and currents, and which determines the voltage induced in an object due to the magnetic field effect. These quantities are related by equation (5-7):

$$H = \frac{B}{\mu} = \frac{1}{\mu} \times \frac{\partial M}{\partial r} \tag{5-7}$$

The voltage induced in a parallel object is given by

$$V_a = j\,\omega\, M \tag{5-8}$$

Where V_a is volts per meter of the object's length, $\omega = 2\,\pi f$, f being the frequency, H_z, of the magnetic field, and M is the magnetic potential of the object "a."[13]

Comparisons between magnetic field induced current and electric field induced current in the human body have been made using a circular cross-sectional model for body elements.[14] These comparisons show that for an electric field of 10 kV/m and a magnetic field of 0.5 G, values chosen as representing maximums known to exist under high voltage power lines, that the electric induced currents are at least ten times greater than the magnetic induced currents. To conclude from this fact that magnetic fields are of less biological importance to humans and animals than electric fields due to power lines is at best premature at this time and at worst non-scientific.

As will later be discussed in this book, the preponderance of experimental evidence that has been developed during the past twenty-five years indicates that electromagnetic field health effects on biological systems may not be insignificant in our environment and that we should proceed with caution in respect to the exposure of humans to EMFs at all frequencies. As long ago as 1972, Robert

O. Becker, M.D.[15] wrote that, "the concept that electromagnetic forces might have any effect upon living organisms—other than the thermal effect due to Joule heating—was for many years rejected by the organized biomedical community. But under the weight of experimental evidence this attitude is changing."

Effects of Ionizing Electromagnetic Radiation

Ionizing radiation is any electromagnetic (EM) or particulate radiation capable of ionizing matter, directly or indirectly. Ionization is the process or the result of any process whereby a neutral atom or molecule acquires a positive or negative charge.

An example or ionizing radiation is found in the photon energy emitted by the sun as this electromagnetic energy passes through air. Another example of ionizing radiation is the emission of both ionizing electromagnetic energy in the form of Gamma rays and Alpha and Beta particles from radioactive materials.

Most Alpha particles are emitted at energy states which correspond to approximately 3,000 megahertz (MHz), or at the lower range of the SHF radio band (just above the UHF band). Gamma rays start at about 10^{15} MHz and hard X-rays can be as high in frequency as Gamma rays.[16] Microwave radiation refers rather loosely to EM energy propagated at frequencies from about 1,000 MHz* upwards[16,17] to about 40 x 10^3 MHz. Ionizing radiation starts at 7.9 x 10^8 MHz or for wavelengths less than 3800 Å (Angstroms†).

Ionizing radiation (IR) as well as particle radiation has been recognized since the late 19th century as a potential health hazard to biological systems, specifically humans and animals. From the beginning, research biomedical scientists identified the dependence of dosage and duration of IR exposure in assessing the risk of hazardous health effects.[18]

Measures of ionizing radiation have evolved from the early units of the rad and rem, the rad being an absorbed radiation unit

---

* 1000 MHz = 10^3 x 10^6 Hz = 10^9 Hz = 1 gigahertz (GHz).

† 1 Å = 10^{-10} meters = 10^{-8} cm. Note that wavelength in meters is given by 300 divided by frequency in MHz.

equivalent to 100 ergs per gram of absorber,[19] and the rem which is equivalent to the rad in units but was introduced later to account for the biological effects of frequency and particulate radiation that differ from X-rays and gamma rays. The SI unit currently used is the sievert (Sv) which is a dose equivalent that takes into account quality factor (Q) and the product of other multiplying factors (N) stipulated by the International Commission on Radiological Protection. One sievert is equal to 100 rems and is the dose equivalent of one joule per kilogram of exposed material.[20]

One sievert is generally considered the threshold above which exposure to IR causes greatly increasing damage to the body, including radiation sickness and radiation dermatitis, cataracts or failure of various organs. Some of the health effects of IR occur many years after exposure, typically five to fifteen years for leukemia, and 40 years or longer for skin, lung, breast and other cancers.[21] The International Commission on Radiological Protection has determined that the total risk of death from IR exposure is about 1% of the exposed population per sievert of absorbed radiation.

Effects of Non-Ionizing Electromagnetic Radiation

The relationship between non-ionizing electromagnetic radiation (EMFs) and humans was not recognized as a quantitative problem until approximately fifty years after d'Arsonval established the use of electromagnetic radiation in medicine in 1890. It was not until 1941 that a scientific team conceived and developed the first true radio frequency dosimetry by quantifying the temperature rise in exposed tissues in terms of a volume-normalized energy rate of absorption. This early work was surprisingly close to the presently used mass-normalized specific absorption rate (SAR) expressed in watts per kilogram.[22]

From the mid through the late 1960s laboratory bioeffects research by Guy,[23,24] Justesen[25] and Gandhi[26] resulted in improved understanding of animal orientation with respect to incident electromagnetic waves and their energy absorption. Other environmental factors, such as temperature and humidity, in addition to dosimetric quantities were found to be of importance.[27] In 1979 Durney et al. developed an empirical formula for broad-band SAR calculations applicable to humans and animals.[28]

Experimental work with microwave radiation was believed as recently as 1961 to have shown "that the chief effect of microwave energy on living tissue is to produce heating."[29] However, there was then and is much more so today a growing body of research that suggests many more complex interactions by the human body with microwave EMFs.

W. Ross Adey and others have since the early 1960s investigated cerebral tissue interactions with environmental electromagnetic fields.[30] The research by Adey and his associates at the UCLA Brain Institute and Department of Anatomy has clearly shown altered electro-chemical effects on the central nervous system due to weak VHF (30-300 MHz) fields, amplitude modulated at brain wave frequencies.[31] Other researchers have observed effects of weak fields at 1-100 MHz radio frequencies on blood cell chaining.[32]

Russian scientists have reported central nervous system (CNS) effects at very low microwave power levels, below 10 mW/cm^2.[33-35] As a result of this research the Soviet government in the mid-1970s established safe exposure levels for microwave radiation 1000 times lower than the U.S. level at that time of 10 mW/cm^2.

The U.S. safety standards for human exposure to microwave energy have evolved from the 10 mW/cm^2 value considered safe for continuous exposure at all frequencies from 10 MHz to 100 GHz, based upon thermal considerations of maximum allowable core temperature rise in the body, to a frequency dependent standard which thermally accounts for the increased absorption of the body particularly near body resonance frequencies of approximately 30 to 300 MHz.[36] The latest recommended standards are specified in the current version of ANSI C95.[37]

ANSI C95.1-1991, IEEE Standard for Safety Levels With Respect to Human Exposure to Radio Frequency Electromagnetic Fields, 3 kHz to 300 GHz includes recommendations to prevent general EMF harmful effects to humans but is not intended to apply to the exposure of medical patients as practiced by the healing arts.[38]

The current recommended standards are based upon the concept of Maximum Permissible Exposure (MPE) in two classes of environments: controlled and uncontrolled. An uncontrolled environment is a location where exposed individuals have no knowledge or control of their exposure. Such areas as homes or workplaces where there are no expectations of exposure beyond those specified in C95.1-1991 Table 2, p. 15, are classified as uncontrolled environments. A controlled environment is, on the other hand, a location where exposed individuals are placed as a result of their employment and where the individual is aware of the potential for such exposure. Such areas of exposure are limited by C95.1-1991, Table 1, p. 13.

In brief summary of the recommended MPE's for these environments the following limits are stated:

1. The maximum MPE for all frequencies is an E field strength of 614 V/m and the minimum MPE is 27.5 V/m.
2. The maximum MPE for all frequencies is an H field strength of 163 A/m (2.05 G) and the minimum MPE is 0.0729 A/m (916 μG).
3. The maximum averaging time* for either E or H field strengths as previously stated is 30 minutes and the minimum is 6 minutes for frequencies up to 15 GHz.
4. The maximum MPE's for both E and H fields are specified for the lowest frequencies, (*i.e.*, 3 kHz to 100 kHz).

C95.1-1991 has also specified recommendations for induced and contact radio frequency currents (3 kHz to 100 MHz) in humans which basically are as follows for uncontrolled environments:

Frequency Range	Max. Current (mA) Through each foot	Contact
0.003–0.1 MHz	450f	450f
0.1–100 MHz	45	45
where f = frequency in MHz		

* Averaging time is a concept designed to determine whether an exposure is in compliance with the MPE. Any exposure exceeding the averaging time is termed a continuous exposure and is considered harmful to humans.

The recommendations for controlled environments are basically as follows:

Frequency Range	Max. Current (mA) Through each foot	Contact
0.003–0.1 MHz	1000f	1000f
0.1–100 MHz	100	100
where f = frequency in MHz		

A method for calculation of the total MPE in an environment that includes multiple sources of electromagnetic radiation is presented in Appendix C of this IEEE publication. There is also a section of the standard which recommends procedures for measuring electric and magnetic fields and presents cautionary advisory information to the engineer which is intended to standardize and improve the data gathering accuracy of such field measurements.[39] Included in C95.1–1991 is an extensive bibliography of 60 citations and a list of 321 papers comprising the initial database for the standard.

Effects of Extremely Low Frequency EMFs

Biological effects of electric and magnetic fields at extremely low frequencies (ELF) below about 300 Hz, such as power frequencies of 50 and 60 Hz, have generated scientific curiosity and research since the late 19th century. Prior to this however, the first scientific observations of the relationship between biological systems and man-made electricity were published by Stephen Gray in England in the 1731 Philosophical Transactions in a paper entitled "Experiments concerning electricity." Gray used static electricity to "electrify" a human subject. By the mid-1700s DC electricity was being generated with Leyden jar batteries and the interest in biological experimentation with DC was greatly increased. There were then a number of physicians and unfortunately outright charlatans who were crudely using this new source of energy to treat afflictions in humans.

In 1786 Galvani discovered, during the dissection of a frog, that an electrical force somehow existed within the frog's leg muscles. From that time on physicians and biologists have recognized and investigated the electrical properties of living tissue. It was not until 1941 however when Szent-Gyorgyi, a physician and biochemist, proposed a startling hypothesis that the biological structure of protein was sufficiently structured to function as a crystalline lattice and therefore that semiconduction could exist in living systems.[40] This was a particularly advanced hypothesis considering that Shockley, Bardeen and Brattain would not actually produce a working semiconductor transistor until after World War II and that Shockley did not publish the theory of the p-n junction semiconductor until 1949.[41]

Simultaneously, during WWII, there was a new science of "cybernetics" being developed at MIT and Harvard which deals with communication and control, whether in man or machine. These advanced concepts began to be applied to neurophysiology and to the complex problem of modeling the brain function. It was finally recognized by biologists that the simple nerve impulse or action potential was not the sole basis for all nervous system functions. These radical concepts were further extended to the central nervous system (CNS) and soon investigators began to experiment with electrically influenced growth and regeneration of bone tissue.[42]

The effects of EMFs at ELF on animals have been generally recognized by today's scientific community to include a wide range of biological systems: the central nervous system (CNS),[43] psychological behavior,[44, 45] migration, navigation and communication,[46] melatonin production by the pineal gland,[47] growth and development,[48] and tissue regeneration.[49]

Specifically the question of EMF effects at power line frequencies on humans has been the subject of scientific research since the 1960's when the U.S. Navy began project Sanguine, an extremely large antenna proposed for communication with submarines during underwater operations. Sanguine was to utilize frequencies in the same range as commercial a.c. power lines and create similar magnetic field intensities, but to have an electric field about one-millionth the strength of a typical transmission line. The Navy, in compliance with the National Environmental Policy Act, initiated experimental studies at various universities and medical research laboratories to assess the effect of such EMFs. Ultimately many of these studies reported positive effects in which the fields produced specific changes in the biological system being studied.

Also during the 1960s, in the U.S.S.R., studies were being conducted on overhead high voltage lines and their biological effects. The Russian EMF work included epidemiological studies which led the Soviet Ministry of Electricity in 1970 to establish rules limiting the duration of exposure to EMFs for electrical workers.[50, 51] Continuing Soviet research resulted in the adoption of additional EMF exposure regulations in 1975.[52]

In the U.S. research on the EMF effects of overhead power lines was begun in 1967 by the Johns Hopkins studies which were sponsored by the electrical utility industry. Although the 1967 studies found reduced sperm counts in linemen and stunted growth in the male offspring of exposed mice, these animal studies were not continued.[53] The studies on linemen were continued and eventually reported as negative.[54] In 1972 the Electric Power Research Institute (EPRI) was founded and it took over almost all responsibility for utility industry and DOE sponsored EMF research. EPRI has continued to sponsor research in biological effects, but only recently has begun to acknowledge in its published EMF effects updates that there may be important human health hazards associated with such exposure.[55]

The independent epidemiological work of Nancy Wertheimer and Ed Leeper at the University of Colorado in 1976-77 was the first study to indicate a possible connection between EMF exposure and human cancer.[56] This study, which was published in 1979 in the American Journal of Epidemiology, was almost immediately refuted by a Rhode Island study by Fulton published in the same journal in 1980.[57] Some in the utility industry were also quick to criticize the Wertheimer and Leeper research and despite a detailed letter to the editor regarding the Rhode Island study which gives several explanations for the different results, the industry generally continued to ignore the Colorado study in favor of the negative result study.[58] In 1982, a Swedish epidemiological study confirmed the Colorado study.[59] In 1987, the IEEE reported that Leonard Sagan, program manager for radiation studies at EPRI, called a New York State study appearing to link exposure to EMFs with leukemia and other types of cancer in children as "well conceived and conducted."[60]

Additional epidemiological studies have been conducted in Los Angeles by a team of researchers from the University of Southern California School of Medicine which seem to confirm the Colorado study and importantly use the wire code classification method developed by Wertheimer and Leeper.[61] It is interesting that many utilities, as well as EPRI, identified the wire code method of magnetic field characterization as being non-scientific when it first critiqued the study by Wertheimer & Leeper in 1979.

The current posture at EPRI is to seriously consider research that examines epidemiological work suggesting a possible link between EMF exposure and childhood leukemia as well as the results of recent laboratory animal experiments which suggest a possible link between the suppression of melatonin production and the development of cancer. EPRI indicates that it plans to sponsor laboratory studies to investigate the exposure of humans to magnetic fields and the suppression of melatonin synthesis. Another area of EPRI indicated research sponsorship is the human exposure to magnetic fields in the work environment.[62]

The earth has a natural electromagnetic environment that varies in frequency from 0 Hz (DC) to several thousand kilohertz (kHz). At 50 to 60 Hz frequencies the earth's background EMF strength is on the order of 10^{-4} V/m (0.0001 volts per meter) and 10^{-8} G (0.00000001 Gauss) for electric and magnetic fields respectively.[63]

The widespread use of electrical energy in our modern world has greatly increased the EMF exposure beyond that of nature. Typically the 60 Hz EMF strengths inside a single family residence in the U.S. will be on the order of 10^{-1} V/m and 10^{-5} G.[64]

Field strengths near residential appliances however can be as high as 300 V/m (20 cm above a rotisserie with a reversed plug) and 1.12 G (3 cm above a 1000 W stove coil).[65] It should be noted that these levels of exposure for humans would usually be very brief and would require extremely close proximity to the appliance. Certain home appliances however can produce high EMFs and are subject to extended human exposure time such as electric blankets and water bed heaters where the fields can be as high as 250 V/m and 1.0 G.

In the previously presented section of this chapter on *Directly Induced Current Flow Within the Human Body,* the effects of overhead power lines (OH PL's) on environmental EMFs were stated. In summary, OH PL's create electric fields in the vicinity of the lines that can be as high as 10 kV/m at ground level[66] and can produce magnetic fields of 0.5 G.[67] These EMFs are considerably greater than the usual strengths found inside a single family residence of 0.1 V/m and 0.00001 G. In fact if a person lived directly under such an OH PL, they would be exposed a significant amount of time to electric and magnetic fields which are 100,000 times and 50,000 times greater, respectively, than the usual EMFs found inside a residence.

The subject of ELF EMF health effects is still in its infancy as far as science is concerned. It has been the author's attempt to explain the current state of understanding that we have in this matter. It is also the author's opinion that we have reached a level of scientific proof concerning the reality of health hazards from EMF radiation at all frequencies sufficient to propose a preliminary stratagem for mitigation of this potential hazard that will currently be effective and practical in minimizing risk.

Several environmental law attorneys have written recent articles on the subject of ELF EMFs and the law.[68, 69, 70] Bruce H. DeBoskey[71] has identified two principal questions that the early EMF research has so far been directed toward:

1. Does electromagnetic radiation (EMR) interact with biological systems in a way that causes effects?
2. If so, are the effects harmful to human health?

As DeBoskey has observed and as has already been cited herein, the generally accepted answer to the first question is yes. The answer to the second question is still subject to scientific debate, however only in specific areas. There can be no question that high frequency EMR, both ionizing and non-ionizing, can be a human health hazard. The biological effects of microwave and ionizing radiation have been generally recognized by scientists and the public for nearly 50 years.

Safety standards have been established by private agencies and governments worldwide that set limits on the exposure of humans to such EMFs. Although the agreement on threshold levels for these safety standards is not uniform, such standards are intended to reasonably minimize the human health hazard of EMR.

There is less general agreement to the answer to the second question as concerns ELF EMF health hazards. However, there is a growing acknowledgement by most researchers that EMR at power line frequencies of 50 to 60 Hz may produce adverse effects upon human life.

What still remain unanswered are the following questions:

3. What are the biological mechanisms that account for ELF EMF interactions in humans?
4. What is the quantified relationship between exposure to such fields and risk to human health?
5. What should be done to reasonably minimize the health risk of exposure to ELF EMFs?

The answer to the third question may require years of research and will undoubtedly bring to light further complex questions. This continuing endeavor is the essence of science.

The fourth question will be resolved by epidemiological studies and laboratory research when such work has been successfully completed to isolate the cause-effect relationship along with the requisite statistical relevance to risk. Extensive debate between scientists and engineers will be required to achieve a quantitation of exposure and risk and the standards setting process implied by the fifth question will occupy the efforts of many knowledgeable people for years.

We should not hesitate to anticipate an eventual resolution of these questions and to recommend even at this time preliminary steps to minimize the health risk to the public from exposure to man made ELF EMFs.

It has already been stated with respect to household electrical appliances that most high EMFs are very brief in duration and usually require very close proximity to the source (*see* p. 92 of this chapter). The exceptions to these conditions are found in electric blankets and pads, water bed heaters, and hair dryers. These appliances are likely to be used for extended periods of time, for five minutes to several hours, and represent the most likely hazard producing sources of EMFs in common use by the general public. A user can totally eliminate the hazard from electric blankets and water bed heaters by turning such devices off immediately before use, (*i.e.*, allowing the stored energy in the blanket or bed heater mass to supply heat for several hours with the power off). Electric heating pads are probably not usable when turned off since their mass isn't very large and hair dryers are not functional when the power is off, therefore these appliances represent an uncontrollable EMF hazard unless manufacturers of such appliances provide future designs that incorporate field emission control or shielding.

Sources of high EMF emission in the workplace include such devices as electric furnaces, welders, and large motor operated machines. The highest reported occupational exposures at 60 Hz are 10 kV/m and 100 G.[72] These sources should be shielded to reduce their electric and magnetic field emission or the workers should be equipped with exposure reducing clothing and apparel to minimize the hazard. Product design for emission field control is ultimately the most effective EMF hazard prevention method for human protection, just as double insulated and ground-fault protected tools have been proven the most effective electrical shock prevention methods for both the general public and the work force.

More than 1,100,000 circuit miles (one way distance from source to load, not one way distance times the number of conductors) of overhead high voltage commercial electric power lines now exist throughout the U.S.* More than 165,000 miles of these circuits operate at voltages from 143 kV to 765 kV (transmission lines) and over 475,000 miles operate at between 22 kV and 143 kV. Research transmission lines have been operated as high as 1,500 kV. All commercial lines are operating to transfer energy and exhibit associated EMFs that are inherent to the transfer process (*see* pp. 79–80 of this chapter).

In order to reduce the existing and future hazard to the public from exposure to power line EMFs, several possible solutions have been suggested:

1. Provide space separation by designating power line corridors whose widths are sufficient to assure that the EMFs from the OH PL's are safe in areas regularly occupied by the public.

---

* Using data from the *EEI*[73] for circuit miles of OH electric lines from 22 kV to 143 kV, the author estimated at least the same number of circuit miles to exist from 5 kV to 22 kV.

2. Utilize natural and man-made shielding to reduce the EMF strength near the lines.
3. Place existing OH lines underground to achieve earth shielding of the EMFs.
4. Utilize new engineering designs for OH PL's that will reduce the EMFs more rapidly as the distance from the line increases.

Each of these solutions has both desirable and undesirable consequences. The solutions all have inherent costs which exceed the present cost of design and construction utilized by the electric utilities. Ultimately the public will pay these increased costs either through energy rates or by taxation if government involvement impacts the public and privately owned utility companies.

The first solution, increasing the width of power line corridors, is effective when the land acquisition costs are not prohibitive. Usually this will be the case only for new lines routed through undeveloped areas, however careful attention to the human use of property immediately adjacent to OH PL's can yield solutions that do not require the general widening of existing corridors.

The second solution is effective when natural terrain and vegetation are available to provide shielding or where man-made structures such as all metal buildings are nearby or will be constructed near the lines.

The third solution, undergrounding the lines, is generally very effective as regards reducing EMF effects, but it is usually much more expensive than OH line construction and is not technically nor economically feasible for extremely high voltage transmission lines or where long distances are involved.

The last solution is chiefly applicable to new line construction and obviously dependent upon research and innovation in electrical power engineering for its effectiveness. Ultimately, the electric utility industry has in the past and will in the future continue to adapt its methods of design, construction and operation of power line facilities to meet the public need for safe and reliable electric energy.

Until the relationship between exposure to ELF EMFs and risk has been quantified, it is the author's opinion that the E and M field strengths at the edge of right-of-way (ROW) for OH electric power lines should be less than three orders of magnitude (1,000) times the presently acknowledged average field strengths inside a single family residence, (*i.e.*, EMFs at the edge of ROW that do not exceed 100 V/m and 10 mG). These limits should also be imposed upon household and work place electrical devices which are likely to require human use and subsequent exposure for extended time periods.

Chapter Notes

1. Ernst Weber and Frederik Nebeker, *The Evolution of Electrical Engineering, A Personal Perspective,* IEEE Press, 1994, pp. 32-33.

2. William M. Mazer, D.E.E., P.E., *Electrical Accident Investigation Handbook,* Electrodata Inc., 1987, Vol. II, Ch. VII.

3. Theodore Bernstein, Ph.D., *Theories of the causes of death from electricity in the late nineteenth century,* University of Wisconsin, Engineering Experiment Station, Reprint No. 1863, p. 267.

4. C.F. Dalziel, "Reevaluation of lethal electric currents," *IEEE Transactions on Industry and General Applications,* IGA-4, No. 5, Sept./Oct. 1968, pp. 467-476.

5. Theodore Bernstein, Ph.D., "Effects of Electricity and Lightning on Man and Animals," ASTM copyright 1973, authorized reprint from the *Journal of Forensic Sciences,* Vol. 18, No. 1, p. 8.

6. C.F. Dalziel, "A study of the hazards of impulse currents," *AIEE Transactions,* 72, Part III, 1953, pp. 1032-1053.

7. Theodore Bernstein, Ph.D., *Theories of the causes of death from electricity in the late nineteenth century,* op. cit., p. 9.

8. Ibid, p.10.

9. William M. Mazer, D.E.E., P.E., *Electrical Accident Investigation Handbook,* op. cit., Sect. 7.4.2.0.

10. *EPRI Transmission Line Reference Book,* 1982. p. 348.

11. ibid, pp. 355-356.

12. ibid, p. 372.

13. ibid, p. 365.

14. ibid, p. 380.

15. Robert O. Becker M.D., "Electromagnetic Forces and Life Processes," *Technology Review,* Vol. 75, No. 2, Dec. 1972, MIT Alumni Assoc.

16. *CRC Handbook of Chemistry and Physics,* David R Lide Ed. in Chief, CRC Press 73rd Ed., 1992-93. p. 10-298.

17. *IEEE Standard Dictionary of Electrical and Electronics Terms,* ANSI/IEEE Std. 100-1984. p. 547.

18. A.W. Guy "Quantitation of Electromagnetic Fields in Biological Systems," *Biological Effects in Electromagnetic Radiation,* IEEE Press, Ed. by John M. Osepchuck, 1983. pp. 1-5.

19. *IEEE Standard Dictionary of Electrical and Electronics Terms,* op. cit., p. 716.

20. Ibid, p. 836.

21. The American Medical Association, *Home Medical Encyclopedia,* Random House, 1989. p. 846.

22. A.W. Guy *"Quantitation of Electromagnetic Fields in Biological Systems,"* op. cit., p. 1.

23. A.W. Guy and J.F. Lehmann "On determination of an optimum microwave diathermy frequency for a direct contact applicator," *IEEE Trans. Bio-Med. Eng.,* Vol. BME-13, 1966. pp. 76-87.

24. A.W. Guy, J.F. Lehmann J.A. McDougall and C.C. Sorensen "Studies on therapeutic heating by electromagnetic energy," *Thermal Problems in Biotechnology,* ASME, 1968. pp. 26-45.

25. D.R. Justesen, and N.W. King, "Behavioral effects of low level microwave irradiation in the closed space situation," *Proc. Biological Effects and Health Implications of Microwave Radiation Symp.,* Richmond, VA, (PB 193898), Sept. 1969.

26. O.P. Gandhi "Frequency and orientation effect on whole animal absorption of electromagnetic waves," *IEEE Trans. Bio-Med. Eng.,* Vol. BME-22, 1975. PP. 536-542.

27. W.W. Mumford "Heat stress due to RF radiation," *Proc. IEEE,* Vol. 57, 1969. pp. 171-178.

28. C.H. Durney M.F. Iskander H. Massoudi and C.C. Johnson "An Empirical Formula for Broad-Band SAR Calculations of Prolate Spheroidal Models of Humans and Animals," *IEEE Trans. Microwave Tech.,* Vol. MTT-27, 1979. pp. 758-763.

29. Mumford, W.W., "Some Technical Aspects of Microwave Radiation Hazards," *Proc. IRE,* Vol. 49, 1961. p. 429.

30. W.R. Adey "Evidence for Cooperative Mechanisms in the Susceptibility of Cerebral Tissue to Environmental and Intrinsic Electric Fields," *Biological Effects of Electromagnetic Radiation,* J.M. Osepchuk ed., IEEE Press, 1983. pp. 108-116.

31. S.M. Bawin L.K. Kaczmarek and W.R. Adey "Effects of Modulated VHF Fields on the Central Nervous System," *Ann. N.Y. Acad. Sci.,* Vol. 247, 1975. pp. 74-81.

32. C.C. Johnson, and A.W. Guy "Non-Ionizing Electromagnetic Wave Effects in Biological Materials and Systems," *Proc. IEEE,* Vol. 60, 1972. p. 693.

33. Yu A. Kholodov *The Effect of an Electromagnetic Field on the Central Nervous System,* Libr. Cong., Washington, D.C., ASTIA Doc. 284 123, 1962.

34. Yu A. Kholodov *Effect of a UHF Electromagnetic Field On The Electrical Activity of a Neuronally Isolated Region of the Cerebral Cortex,* Libr. Cong., Washington, D.C., ATF P 65-68, 1964.

35. Yu A. Kholodov *The Effect of Electromagnetic and Magnetic Fields on the Central Nervous System,* NASA Technical Translation, TTF-465, 1967.

36. A.W. Guy, "Non-Ionizing Radiation Dosimetry and Interaction," *Biological Effects of Electromagnetic Radiation,* J.M. Osepchuk ed., IEEE Press, 1983. pp. 530-531.

37. *ANSI/IEEE C95.1-1991,* "IEEE Standard for Safety Levels with Respect to Human Exposure to Radio Frequency Electromagnetic Fields, 3 kHz to 300 GHz," IEEE, 1992.

38. ibid, p. 9.

39. ibid, pp. 29-32.

40. R.O. Becker and Andrew A. Marino, *Electromagnetism & Life,* State University of NY Press, 1982. Ch. 1, p. 21.

41. W. Shockley *Bell System Tech. Jour.,* Vol. 28, 1949. p. 435.

42. R.O. Becker and Andrew A. Marino, *Electromagnetism & Life,* op. cit., pp. 40-56.

43. H.A. Hansson "Purkinje Nerve Cell Changes Caused by Electric Fields - Ultrastructural Studies on Long-Term Effects on Rabbits," *Medical Biology,* 59:103-110, 1981.

44. W.R. Gibson and W.F. Moroney "The Effects of Extremely Low Frequency Magnetic Fields on Human Performance," *AD A005898, NAMRL-1195,* Pensacola, FL; Naval Aerospace Medical Research Laboratory, 1974.

45. J.R. Hamer, "Effects of Low-level Low-frequency Electric Fields on Human Reaction Time," *Commun. Behavioral Biology 2,* Part A, 1968, p. 217.

46. R.O. Becker and Andrew A. Marino, *Electromagnetism & Life,* op. cit., pp. 70-74.

47. Taylor Moore "Sharpening the Focus on EMF Research," *IEEE Power Engineering Review,* Vol. 12, No. 8, Aug. 1992, p. 8.

48. Jose M.R. Delgado J. Leal Jose L. Monteagudo, and Manuel G. Gracia "Embryological Changes Induced by Weak, Extremely Low-frequency, Electromagnetic Fields," *Journal Anat. (Gr. Brit.),* 1982, pp. 533-550.

49. R.O. Becker and J.A. Spadaro "Electrical Stimulation of Partial Limb Regeneration in Mammals," *Bull. N.Y. Acad. Med.,* 48:627.

50. N.A. Solov'ev "The Biological Effect of the Electric Component of a Low Frequency Electromagnetic Field (Literature Survey)," *Medical Technology News, Moscow,* VNIIMIO, No. 5, 1962, pp. 86-94.

Yu I. Novitskiy Z.V. Gordon A.S. Presman and Yu A. Kholodov "Radio Frequencies and Microwaves: Magnetic and Electric Fields," *Foundations of Space Biology and Medicine,* Vol. 2, Part 1, Academy of Sciences, USSR, Moscow, 1970. (NASA

Washington, D.C., Technical Translation TT-F-14, 021, Nov. 1971).

Yu L. Dumanskiy V. Ya Akimenko and Ye V. Prokhvatilo "Biological Effects of Low Frequency and Static Electromagnetic Fields," *USSR Ministry of Health*, Moscow, 1977.

51. V.P. Korobkova Yu A. Morozov M.D. Stolarov and Yu A. Yakub "Influence of the Electric Field in 500 and 750 kV Switchyards on Maintenance Staff and Means for Its Protection," *Proceedings of the International Conference on Large High Tension Systems,* Paris, 1972.

52. Yu I. Lyskov and Yu S. Emma "Electrical Field as a Parameter Considered in Designing Electric Power Transmission of 750-1150 kV: The Measuring Methods, the Design Practices, and Direction of Further Research," *Proceedings of the Symposium of EHV Power Transmission, Joint American-Soviet Committee on Cooperation in the Field of Energy,* Washington, DC, Feb. 1975.

53. Andrew A. Marino, and Robert O Becker, ., "High Voltage Lines, Hazards at a Distance," *Environment,* Vol. 20, No. 9, Nov. 1978.

54. M.L. Singewald O.P. Langworthy and W.B. Kouwenhoven "Medical Follow-Up Study of High Voltage Linemen Working in AC Electric Fields," *IEEE Trans. on Power Apparatus and Systems,* P.A.S. 92:1307-1309, 1973.

55. Bruce H. DeBoskey "The Killing Fields: Electromagnetic Radiation and Cancer," *TRIAL,* Sept. 1993, pp. 54-58.

56. N. Wertheimer, and E. Leeper "Electrical Wiring Configurations and Childhood Cancer," *Am. J. Epidemiol.,* 109:273, 1979.

57. J.P. Fulton, S. Cobb L. Preble L. Leone, and E. Forman "Electrical Wiring Configurations and Childhood Leukemia in Rhode Island," *Am. J. Epidemiol.,* 111:292, 1980.

58. N. Wertheimer and E. Leeper "Electrical Wiring Configurations and Childhood Leukemia in Rhode Island," *Am. J. Epidemiol.,* 111:461, 1980.

59. L. Tomenius L. Hellstrom and B. Enander "Electrical Constructions and 50 Hz Magnetic Field at the Dwellings of Tumor Cases (0-18 years of age) in the County of Stockholm," *Symposium on Occupational Health and Safety in Mining and Tunneling,* Prague, June 21-25, 1982.

60. *IEEE Spectrum,* "Spectral Lines," Vol. 24, No. 9, Sept. 1987, p. 25.

61. Taylor Moore "Sharpening the Focus on EMF Research," op. cit. 16., p. 7.

62. Ibid, pp. 8-9.

63. A.R. Sheppard and M. Eisenbud *Biological Effects of Electric and Magnetic Fields of Extremely Low Frequency,* New York University Press, 1977, p. 2-1.

64. EPRI, "Transmission Line Reference Book," *EPRI,* 1982, p. 348.

65. A.R. Sheppard and M. Eisenbud *Biological Effects of Electric and Magnetic Fields of Extremely Low Frequency,* op. cit., pp. 3-3,4.

66. EPRI, "Transmission Line Reference Book," op. cit., p. 347.

67. EPRI, "Transmission Line Reference Book," op. cit., pp. 342-343.

68. B.H. DeBoskey "The Killing Fields: Electromagnetic Radiation and Cancer," *Trial,* Sept. 1993. 99. 54-58.

69. J.P. Foote "The Threat of EMF: Fact or Fiction," *The Legal Investigator,* Nov. 1993. pp. 35-38.

70. M.E. Withey "Proving Causation in EMF Based Claims: It's the Method, Not the Madness!," *CTLA Forum,* Dec. 1993.

71. B.H. DeBoskey "The Killing Fields: Electromagnetic Radiation and Cancer," op. cit., p. 54.

72. "Proceedings of the Scientific Workshop on the Health Effects of Electric and Magnetic Fields on Workers," *NTIS,* PB92-131721, U.S. Dept. of Commerce, Jan. 1991. p. 49.

73. *EEI Statistical Yearbook of the Electric Utility Industry,* 1992. p. 97.

Chapter 6

Case Studies

Introduction

Since 1957 and for the intervening thirty-seven years, the author has practiced electrical engineering. Since 1968 he has provided forensic engineering services to the legal profession. During the years since 1968, the forensic engineering portion of this practice has steadily grown such that today it occupies more than 90% of the author's professional effort.

During this engineer's work in forensic electrical engineering, by far the greatest number of cases have dealt with electrocutions, 50% of all cases. Fires of possible electrical origin have constituted the second largest class of cases, 30%, and the remaining 20% of the cases have involved in descending order, product failures, electromagnetic fields, illumination, construction, lightning, water damage to electrical equipment, and non-lethal damage due to overhead line contact.

In the early years most of this forensic engineering work was done for plaintiff's attorneys, but this has changed so that today the cases are about equally divided between plaintiffs and defendants.

There has also been a sizable number of forensic investigations, conducted for insurance companies, utilities, etc. which were performed for unidentified legal positions (*i.e.*, neither a plaintiff nor a defendant since no law suit was contemplated).

In the following case studies, the names of litigants are omitted and generic terms will be used to describe locations, products and features so as to avoid unwarranted exposure of parties to the litigations or actions involved.

6.01 A Case of Power Line Electrocution by a Rabbit

In 1982 a farmer and his two young sons were working in a field which utilized thirty foot long aluminum irrigation pipes that could be coupled together to selectively carry water to the plant rows. One of the boys saw a rabbit run into a pipe section, and he called for his father to raise the pipe from one end while he and his brother waited at the low end of the pipe for the rabbit to slide out. As the man walked the pipe upward to a greater and greater height, the pipe made contact with an overhead 13 kV power line conductor that ran beside the field inside the fence line. The electrical current which resulted from the 7,200 V to ground contact killed the man and severely injured both boys. The attorney who represented the decedent's wife and boys contacted this engineer for an opinion and investigation of liability against the electric utility which owned the power line.

In this case the 1977 NESC was in effect, based upon the date of construction of the power line, and this edition of the code had been adopted by the state utility commission at the time. Sec. 232, Table 232-1, of the code requires a minimum clearance above grade for this class of line of 20 feet. The actual clearance of the subject line was 22 ft. 4 in. at point of contact, and the line was 21 ft. 6 in. above grade at its minimum clearance point in the accident span. The code minimum of 20 feet is for a classification of surface underneath the conductors that is "other land traversed by vehicles such as cultivated, grazing, forest, orchard, etc."

The accompanying footnote reads:

> "These clearances are for land cultivated or traversed by vehicles and equipment whose overall operating height is less than 14 ft."

Although it would appear that the electric utility was in compliance with the NESC based upon Sec. 232, there are also more general sections of the code such as Sec. 210 which states "All electric supply and communication lines and equipment shall be of suitable design and construction for the service and conditions under which they are to be operated," and Sec. 211 which states "All electric supply and communication lines and equipment shall be installed and maintained so as to reduce hazards to life as far as is practical."

An investigation into the operation of this farm and others in the vicinity of the subject power line showed that there was widespread use of the 30 ft. long aluminum pipes for irrigation and that such use had been common for more than fifteen years prior to this accident. It was also determined that the pipes, which were about 6 inches in diameter and weigh about 100 pounds each, are easily stood on end by walking the pipe up from a horizontal to a vertical position as the decedent was doing.

Based upon the reasonable knowledge that an electric utility which operates in such a rural agricultural area would have of the potential danger from OH line contact by this equipment, it was the engineer's opinion that the utility was in violation of Sec. 210 and Sec. 211 of the NESC by constructing and operating the 13 kV OH power line inside the irrigated field fence at less than 30 ft. above grade. The specific use of the land where 30 ft. long metallic pipes were used in large numbers along with the ease in standing such pipes upright made it incumbent, in the engineer's opinion, upon the electric utility to construct its open conductor OH high voltage power lines in a manner that would be practical and reduce the hazard to life from line contact with such equipment.

The defendant's attorney deposed the plaintiff's forensic engineer in this case, and the case was settled before trial began. It is interesting to observe that the NESC (see Chapter 3, pgs. 8-10), since the code revision of 1981, has changed Rule 210 and has deleted Rule 211. A less specific rule related to reducing hazards is now found in Sec. 012 which states "All electric supply and communication lines and equipment shall be designed, operated, constructed and maintained to meet the requirements of these rules. For all particulars not specified in these rules, construction, operation, and maintenance should be done in accordance with good practice for the given local conditions." An important change in this wording is inherent in the term "should" instead of "shall," the latter meaning a mandatory requirement and the term "should" meaning a provision that is "normally and generally practical for the specified condition."[1]

The 1990 NESC further reduces the effectiveness of Rule 012 by stating that, "Where the word 'should' is used, it is recognized that, in certain instances, additional local conditions not specified herein may make these provisions impractical. When this occurs, the difference in conditions shall be appropriately recognized and Rule 012 shall be met."

If the question of what happened to the rabbit has occurred to the reader, it is the author's scientific conclusion that the rabbit survived.

6.02 A Case of the Good Samaritan Fire

In June 1979 a home owner was in the process of having his air conditioning system repaired by a licensed air conditioning serviceman. As an adjunct to the service call, the man asked for the serviceman's help in locating an electrical problem at the residence. It seemed that the receptacles in his living room which had table lamps plugged into them and the front porch lights were not functional. After locating the correct interior panelboard, the serviceman found a tripped 15 Amp circuit breaker. He demonstrated to the owner the proper procedure for resetting the breaker; however, the breaker immediately tripped off as he reset it. At this time the serviceman instructed the owner to wait at the panelboard while he returned to the living room area in order to observe any electrical activity before he called for the owner to reset the breaker one more time.

When the owner reset the breaker the second time, the serviceman heard a loud "pop" and saw a "flash" from the side of the living room picture window near the floor. The hanging drapery beside this window then immediately burst into flame. The fire spread was extremely rapid and resulted in both localized structural and widespread smoke damage to the residence.

This engineer was contacted by the owner's insurance company three days following the fire and was requested to investigate the scene which at that time had only been disturbed by the rural-metro fire fighters who had responded to the fire. Upon arrival at the residence, this engineer was given a first hand account of the events that preceded the fire from the owner. A careful examination of the debris in the area of the observed fire origin disclosed the metal remnants of a duplex receptacle and lamp cord plug along with two small portions of a phenolic plastic commonly used in electrical molded insulators that form the body of many duplex receptacles.

Two other smoked but otherwise undamaged receptacles were removed from the residence in order to compare their parts with the discovered remnants and to identify the manufacturer of the receptacle from the area of fire origin. By comparing the parts of the two undamaged receptacles with the partial remnant parts from the fire origin, it was possible to determine how much and where the remnant parts had lost material during the fire. Since the metallic parts from these receptacles are brass and copper, which melt at temperatures between 930° C and 1083° C, it is unlikely that a fire of the type which occurred in this incident would create temperatures sufficient to destroy any of these metallic parts.

However, an electrical arc cannot only melt and vaporize such metals but along with the eyewitness account of a "pop" and a "flash," this evidence suggests strongly that an electrical arc did indeed occur to initiate the fire. There was also evidence of electrical arcing present as globular metal deposited and fused into one of the remnant receptacle terminal clips.

At this point in the investigation, it was apparent that an arc had occurred in the receptacle which had a lamp cord plug inserted in it and that the super-heated metal particles which exploded from that arcing process had ignited the draperies which were hanging directly in front of the receptacle. Several years prior to this case, this engineer had been given a receptacle and plug which an engineer friend had recovered from his own kitchen following an arcing incident which startled both him and his wife. I had examined this receptacle and plug at that time and had determined that the cause of the arcing was a plug which was inserted or which had been displaced in a non-perpendicular manner into the receptacle such that the plug prong had broken through a thin plastic barrier which was designed to separate the plug prong from the grounded metal mounting strap within the receptacle body. The energized side of the receptacle and plug thus came into contact with the ground strap and an electrical arc ensued.

This previous device failure was the basis for this engineer's theory in this fire case that an energized plug prong had come into contact with the grounded metal strap of the receptacle and electrical arcing that created the fire resulted. There is no way to be certain, from the available evidence, whether the receptacle failed because the plug broke through a normal plastic barrier in the device or whether the plastic barrier was defective as manufactured (*i.e.*, a barrier that was missing or formed with inadequate material). In either situation there is a defective product involved because such receptacles must be designed to accept without failure the normal wear and tear and variances that will occur as plugs are inserted or removed. The subject receptacle was, according to the owners, not one that was regularly used, and the table lamp had been continuously plugged into the receptacle since the last time any furniture was moved.

After the engineer's report and conclusions were presented to the insurance company in this case, a subrogation lawsuit was filed against the receptacle manufacturer. Several years later this engineer was deposed by the manufacturer's attorney and the case was subsequently settled out of court. Unfortunately for the air conditioning company that was servicing the residence, a law suit was later brought against the serviceman alleging inadequate technical skill and care for helping the home owner with his electrical problem. Despite this engineer's testimony that it is not unusual nor beneath the standard of care expected of such technicians to reset a breaker more than once during a trouble shooting investigation, the trial against the air conditioning company and its serviceman resulted in an unfavorable jury verdict against the Good Samaritan serviceman.

6.03 A Case of a Utility's Failure to Respond to a Request for Help

In 1975 a contractor was installing approximately 1.4 mi. of eight ft. diameter storm line buried 20 to 25 feet deep along the side of a city street paralleled by a 12 kV overhead power line. During the construction project it was necessary to build a box culvert 30 ft. deep at an intersection, the construction of which required the use of sheet pilings for the excavation. The pile driver frame was nearly 6 ft. higher than the OH PL and the danger was recognized by the contractor as his workers were in the initial process of setting up the derrick and pile driver frame.

The contractor then stopped that portion of the work and telephoned the electric utility to request assistance in assuring the safety of the work near the power line. A field representative from the utility visited the construction site following the contractor's request. The utility field representative observed the OH PL and construction equipment, after which he informed the contractor that "the line cannot be deenergized and you should not bring your equipment within ten feet of the line." After this advice the utility representative left the construction site.

Faced with this "Catch 22" situation, the contractor proceeded to set up the derrick and pile driver as far away as possible from the line and still allow the equipment to drive the sheet pilings. This setup placed the driver frame 2 ft. 3 in. from the closest OH PL conductor. A day or so later, as the pile driver was operating, a laborer on the ground was holding onto a steel cable used to guy the driver frame when the cable moved laterally and made contact with the 7200 V to ground power line. The worker was electrocuted, and his widow filed a law suit against the utility company and the city who was the owner of the drainage line. The contractor/employer was immune from liability under a state worker's compensation law.

An investigation of the case established the following:

1. Prior to starting construction a pre-job conference was held between the contractor, the engineer, the city and the local water, gas, electric and telephone utility representatives. The purpose of the pre-job conference was to inform all parties involved of the project scope and timing and to review any foreseeable difficulties, interferences or safety issues that the project presented.

2. At the pre-job conference the contractor discussed the use of a high profile crane throughout the project to lift and place the 8 ft. diameter concrete drainage pipe sections.

3. The engineer's plans and specifications for the project were furnished to all involved parties prior to the pre-job conference.

4. At the pre-job conference, the electric utility representative cautioned the contractor about the hazard of working around the 138 kV OH transmission line that crossed over the street toward the north end of the project. No offer was made by the electric utility representative to assist the contractor with any problems that might involve OH PL's.

5. The electric utility company's safety manual stresses the problem of crane contacts with its OH lines and emphasizes that the company places a high priority on the safety aspects of prevention of such contacts. The manual strongly recommends that all crane operations near OH PL's use a trained ground spotter and that the high profile equipment be grounded by a bonding conductor to the closest available ground point.
6. During the job a conference with contractor, city inspector and electric utility representative took place to discuss the specific problems foreseen with the box culvert construction.
7. The utility's field representative, who responded to the contractor's telephone call for assistance, was not electrically trained or educated and had only recently been transferred from the utility's gas division to its electrical division.
8. After the accident, at the contractor's expense, the utility built a temporary short line called a "shoofly" around the box culvert construction area and deenergized the conductors that were adjacent to the pile driver.
9. The electric utility line that was involved was a short, less than 2 mi., radial circuit that served no critical loads (*e.g.*, a hospital).

The 1973 NESC, which was in effect at the time of this accident, has three sections that apply to the conduct of the utility in this case:

210. Design and Construction. All electric supply and communication lines and equipment shall be of suitable design and construction for the service and conditions under which they are to be operated.

211. Installation and Maintenance. All electric supply and communication lines and equipment shall be installed and maintained so as to reduce hazards to life as far as practicable.

214. Isolation and Guarding.

A. Current-carrying Parts.

To promote safety to the general public and to employees not authorized to approach conductors and other current- carrying parts of electric supply lines, such parts shall be arranged so as to

> provide adequate clearance from the ground or other space generally accessible, or shall be provided with guards so as to isolate them effectively from accidental contact by such persons.

In deposition and later during trial, this engineer testified that in his opinion "(the electric utility) was definitely negligent in not taking positive action in concert with the contractor to increase the safety of the operation ... and that (the utility) was in violation of the 1973 NESC Sec. 210, 211 and 214.A."

The jury awarded $1.5 million to the family of the deceased laborer. The verdict was upheld upon appeal to the state Supreme Court. In addition, OSHA cited the contractor for violation of state safety regulations and the contractor was fined $4,500.

6.04 A Case of Severe Electrical Shock in the Shower

In 1970 a woman was taking a shower in her bathtub when she was suddenly overcome by a sense of electrical shock apparently coming through the water spray. She passed out and evidently fell in the tub, striking a water control knob as she collapsed which allowed nearly 100% hot water to continue to discharge from the shower nozzle onto her body. Approximately three to five minutes later she regained consciousness and screamed for help. Her husband responded from outside the residence and rushed her to the hospital where she was treated for second and third degree burns over a considerable portion of her body.

This engineer was contacted by the woman's attorney to investigate the electrical shock phenomenon which she apparently experienced and to identify the source of this shock. Early in the case, the husband reported that a 30 Amp fuse supplying one leg of a 230 V water heater was found blown after the incident. The couple also reported that they had been mildly shocked by touching the kitchen sink water faucet prior to the incident.

An engineering investigation of the residential electrical system disclosed the following:

1. The water heater resistance was 2.5 ohms from line 1 of the 230 V supply to the lower water heater element and the cold water pipes.

2. Line 2 to the lower water heater element was open circuited (infinite resistance) to the cold water pipes.
3. The house was grounded at the service entrance via a driven ground rod.
4. The service entrance neutral was bonded to the ground connection but not to the metallic water piping supply system.
5. The metallic water supply line serving the house was electrically isolated from the main utility water lines by an insulating coupling.
6. The resistance between the service entrance neutral and the residence water piping was measured at 7.1 ohms.
7. The resistance from the metal shower drain fitting to the driven ground was measured at 56 ohms.
8. The resistance between the metal shower drain and the bathtub shower head was measured at 120 ohms.
9. The house was wired originally around 1950 using NMC conductors without a separate ground wire and the water heater was installed in 1967 without a separate ground wire.

The lower faulted heater element was removed and examined. An obvious electrical arc had burned near a hole in the element sheath approximately 1/5 th of the distance from the shorted terminal to the open terminal. The other terminal was open circuited to the grounded sheath and to the first terminal. The element was x-rayed to verify the internal condition of the helix resistance wire and its relationship to the sheath. Using x-ray diffraction, emission spectroscopy, and certain chemical analyses of the coating on the heater element, it was determined that the coating was composed almost entirely of carbonates, the greatest portion of which was calcium carbonate in aragonite form. It was determined by serial numbers and manufacturer's records that the heater was manufactured about 2-1/2 years before the incident and it was found that the water heater had been in service about 2 years prior to the accident.

A water sample from the supply to the residence was chemically analyzed for soluble salts, hardness, pH, fluoride and nitrates.

This data was compared with other water analyses from various locations across the country in order to assess the effect of water chemicals on the expected deterioration of the immersion heater element.

An electrical circuit model was formulated to include the measured parameters from the residence electrical and plumbing systems as shown in Fig. 6.1 below.

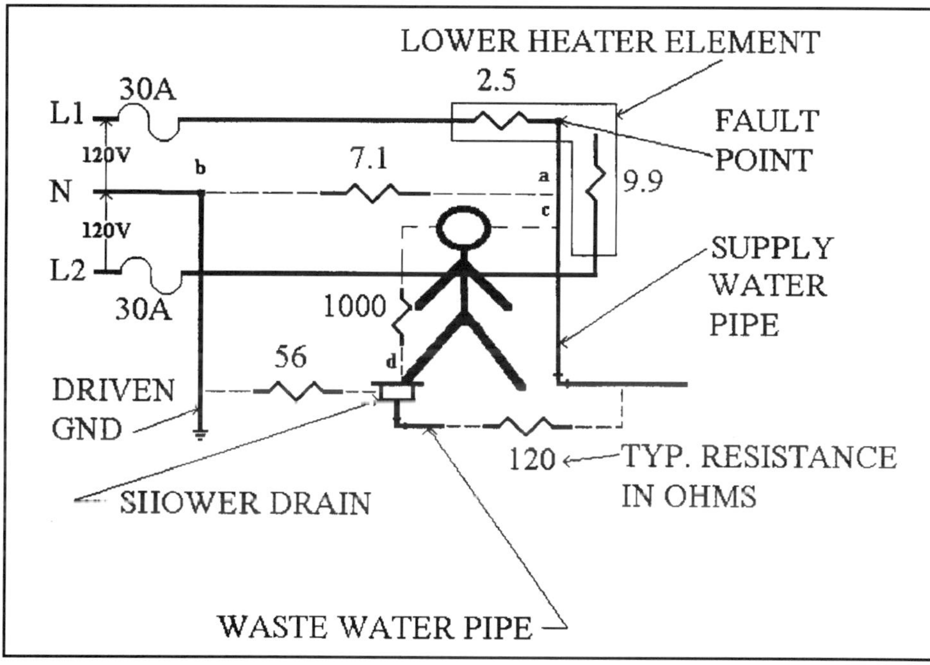

Fig. 6.1 *Water heater case circuit.*

Various voltages were calculated assuming a wet body and water path resistance of 1000 ohms as shown (*i.e.*, resistance from the shower head through the water stream and body to the shower drain). The victim was postulated to have been subject to an initial shock of approximately

$$I = \frac{V_{cd}}{R} = \frac{57.7}{1000} = 57.7\ mA \qquad (6\text{-}1)$$

This shocking current, if continued for a period of more than 10 seconds would likely produce ventricular fibrillation (VF) in a person. Simultaneously, as the woman experienced this shock, there would be a change in the electrical circuit parameters. If the initial current flow through the head and upper torso from the shower water produced a rapid loss of consciousness, then as she fell to the floor of the tub, the path resistance through the water spray and body would increase significantly and the likelihood of VF or electrocution would decrease. Also, as hot water was used from the storage tank, the upper heating element would activate increasing the line current, and as the arcing contact between the failing element helix and sheath lowered its resistance, the circuit current would increase beyond the fuse rating and the fuse would blow. This postulated explanation for the accident agrees in all respects with the discovered facts following the incident.

The plaintiff's attorney, in view of the early failure of an appliance that according to the manufacturer's life cycle testing is expected to last over 15 years, filed suit against the water heater and element manufacturer for a faulty product. The x-ray examination of the water heater element clearly showed that the helix, on both sides of the break and arcing location, was displaced from a centering location within the tube sheath. This strongly suggests a misalignment of the helix as it is loaded and the magnesium oxide insulation is packed around it in the tube.

Such a manufacturing defect would likely pass the final inspections since the manufacturer was relying primarily on hi-pot (high voltage) testing to insure the insulation integrity between the helix and sheath. As the element was used, the heating and cooling of the helix in the region where a slight void or lack of MgO existed would aggravate the defect and would logically result in an early failure such as an arcing fault from the helix to the grounded metal sheath.

This case was ultimately tried. The jury awarded $550,000 to the injured party, which was the largest personal injury award ever given in that state as of that time. Following trial the manufacturer, who had participated with the local REA utility in marketing the electric water heaters within the area, sent a letter to all owners informing them of the necessity to connect a separate ground wire from the appliance to the service entrance grounded neutral.

6.05 A Case of an Attractive Nuisance Substation

A utility substation is an extremely hazardous area for anyone to enter except qualified individuals. Substations are enclosed or semi-enclosed structures which contain the transformers, switches, circuit protective devices, and controls for the conversion and redistribution of electrical energy from transmission or subtransmission voltages, usually above 34 kV, to distribution voltages from 4 to 34 kV. A typical rural area substation is generally a fenced enclosure with overhead power lines entering and perhaps exiting the enclosure. Within the enclosure the lines usually are placed on lowered structures such that open bare conductors are present at levels as low as seven feet eight inches above grade.*

A rural area 69:12 kV substation, approximately 80 ft. by 180 ft., became the object of interest for two 8 yr. old boys. The substation was enclosed by a 7 ft. high chain link fence that had a three strand barbed wire climbing guard at the top. The inside of the substation was relatively open and at the time of the accident to be described herein, an employee of the utility had stored his pickup camper shell inside the locked substation fence.

The boys were attracted to the substation because of a pirate game that they regularly played using the camper shell, which was raised above grade on concrete blocks, that became their ship and hiding spot. They had played several times inside the substation, gaining entrance sometimes by climbing over the fence and wire guards and at other times by slipping under the double gate which had an eroded rain runoff space beneath it. On the day of the accident, one of the boys was using a five foot length of broken auto antenna as a play sword, and while standing next to the substation transformer, he slashed overhead with the sword and made contact with a 69 kV conductor located on the top of an insulated bushing.

Both boys were burned by the arc flash that ensued, one boy was burned severely, but both survived.

This engineer was retained as an expert for the plaintiffs in the subsequent lawsuit. The plaintiffs' attorney had already begun an investigation and had obtained photographs showing the condition of the substation immediately following the accident, copies of the sheriff's report, and hospital records from the incident.

---

* Minimum vertical clearance requirement from the '48 NESC H30, Sec. 114, Table 2, for 600 V between phases. The '90 Code requires 8 ft. 8 in. for this voltage classification.

An engineering investigation revealed that the substation was constructed in 1971. A site inspection by this engineer approximately six months following the incident showed that the substation had an established runoff pattern under the access gates that would exceed twelve inches in height by thirty inches in width unless fill material was used to replace the erosion. Local weather reports showed that the area had received 6-1/2 inches of rain during the 10 days prior to the accident.

Research into the code indicated that the applicable NESC was H30, adopted in 1948. Under H30, Sec. 11 Protective Arrangements of Stations and Substations, Sec. 110 Protective Arrangements of Equipment states "All electric supply equipment shall be of such construction and so installed and maintained as to reduce the life hazard as far as practicable." Also under H30, Sec. 102 General Requirements, Sec. 102 B.2.Storage and Manufacturing Processes, the Code states "They [stations and substations] shall be used neither for the storage of material nor for manufacturing processes . . ., except those materials . . . attendant upon the production or distribution of a supply of electric energy."

Despite the engineer's opinion, expressed during deposition taken by the defendant utility's attorney, that the utility had violated Rule 110 and Rule 102 B.2. of the NESC and further had created an attractive nuisance by allowing an employee to store his camper shell in the substation, the case was tried before a jury. The jury awarded $325,000 to one boy, of which $50,000 was for punitive damages, and $70,000 to the other, of which $50,000 was for punitive damages.

6.06 A Case of Mobile Home Fire due to Faulty Wiring

The attorney for a mobile home manufacturing firm contacted this engineer in 1975 to investigate an alleged electrical fire which severely damaged a mobile home. The subject mobile home had been modified by others following its initial manufacture to allow its use as a real estate office.

The engineering investigation began by an examination of the following materials:

1. The floor plan of the subject mobile home as originally manufactured showing the basic electrical plan.
2. A sales order showing that the mobile home was to be constructed with a 120/240 V 100 Amp electrical service.
3. An invoice of different date confirming item 2.
4. A separate floor plan with hand drawn modifications added.
5. A local fire department run report.
6. A portion of the manufacturer's quality control program showing electrical system testing procedures utilized during and at completion of construction.
7. A report authored by another engineer who had initially investigated the fire.
8. A partial copy of NFPA 501B 1973 Standard of Mobile Homes, Part E Electrical Systems.
9. A portion of wiring evidence approximately two feet in length taken from the subject mobile home.
10. A summary of the deposition of the millwright who modified the original mobile home to allow its use as an office.

Following an examination of this information, the engineer requested and obtained information as to the manufacture and specific product for the circuit breaker load center installed in the mobile home, an inventory of the office equipment in use at the time of the fire, photographs of the fire damaged structure taken shortly after the fire, and photographs from the original manufacturer showing the wiring as typically installed in its mobile homes. Unfortunately, the actual fire damaged mobile home was salvaged about 8 months before this investigation began and, therefore, was not available for inspection by this engineer.

After all the physical evidence, photographs, plans and statements were examined, it was the engineer's conclusion that the fire started in the vicinity of the wiring to a receptacle, specifically at a point where a wire staple had been used to secure the copper NM cable wiring to a wood stud. A distinctive vee fire pattern was present from this location upward, and the depth of char as well as other structural damage from the fire was most severe in this area compared to the other area photographs. This area of origin also agreed with the eyewitness accounts of the fire and the fire department's opinion of an area of origin.

Because of the difference in wiring method used in the area of fire origin compared to the manufacturer's photographs and based upon the hand drawn modifications used to describe the changes made following manufacture, it was concluded that the wiring at the point of fire origin was most likely installed after the product was originally constructed.

This case was eventually submitted for trial by jury. The verdict was entirely in favor of the defense, with no damages awarded to the plaintiffs in this matter.

6.07 A Case of Careless Tree Trimming

In 1986 this engineer was retained by an attorney representing a public utility in defense of a lawsuit involving the death of a tree trimmer. The accident occurred in 1980 when a 49 year old tree trimmer in the process of trimming a palm tree came into contact with an overhead 12 kV distribution line owned by the defendant.

The decedent had apparently used a 29 foot long aluminum ladder to ascend a tall palm tree where he then proceeded to throw a metallic safety chain out from and around the tree. As he did this the chain came into contact with the utility power line which was energized at 7,200 V to ground. The subsequent arcing explosion ignited both the victim's clothes and the tree. The man lost consciousness and while still holding his gasoline chain saw, he fell to the ground.

The chain saw gasoline tank either ruptured upon striking the ground or was ignited during the arcing explosion, and the victim was severely burned across the front of his body while lying on the ground. Although the man was alive and breathing on his own after transport to the hospital, he had sustained third degree burns over 30% percent of his body and second degree burns over 17% of his body. The victim died two months later from the burn injuries.

Plaintiff's attorney sued the utility, the chain saw manufacturer, and the owner's of the residence alleging against the utility that it had, as quoted from the complaint:

a. carelessly, negligently, or intentionally, without due regard for the public or for persons lawfully employed in work taking them close to the wires, maintained wires without insulation.

b. failing to properly inspect and see that the wires do not touch the limbs of the trees, and in permitting the wires to touch or run close to the limbs of trees.

c. failing to warn persons, and particularly Plaintiff's Decedent, of the fact that the wires were carrying electrical current of high voltage and were dangerous.

d. failing to give warning to persons that ordinarily comes [sic] in contact with the tree near which, or through which wires ran, although Defendant power company knew that tree must be trimmed, cut, or treated.

e. failing to take reasonable precautions in the premises or to use due care, in that it failed to install prevent or maintain proper or adequate circuit breakers or fuses or other safety devices to break the electrical circuit when such lines come in contact with Decedent's tools.

f. failing to inspect or to make reasonable inspections of its' lines and equipment so that the dangerous proximity of the high voltage lines to the trees on which decedent was working would have been discovered.

g. failing to have the wires of such high and potential (sic) dangerous voltage of electricity installed underground or installed in proper conduits.

h. failing to warn the public, particularly the decedent of the extremely high voltage being carried by its wires when the voltage was so high that it would arc or jump to the equipment used in cutting trees or trimming trees in the premises.

i. maintaining the high tension electrical wires in such a manner without proper precautions in residential areas as to constitute a public and private nuisance.

The engineer's investigation revealed the following:

1. The subject power line was a minimum of 27 ft. 6 in. above grade in the incident span.
2. The subject power line was 6 ft. 6 in. north of the north face of the incident palm tree.
3. The victim was a part-time tree trimmer who had on previous occasions trimmed the subject tree as well as other trees on the property where the accident occurred.
4. A specific safety clearance is not specified by the NESC for separation of overhead lines and trees, but the provisions of the Code in a more restrictive environment require a minimum horizontal clearance of 5 ft. from the balconies of buildings which are accessible to pedestrians.
5. In all aspects of a detailed review, the existing utility facilities were in conformance with the NESC at the time of the accident and the line was installed, operated and maintained according to reasonable standards of care within the utility industry.

Despite this engineer's affidavit and subsequent statement, that clearly supported the utility, taken by the plaintiff's attorney in deposition, the case was taken to trial before a jury. The jury found no liability on the part of the utility in this case.

6.08 A Case of an Exploding Bus Duct

In July 1982 an outdoor bus duct serving a newspaper production and office facility failed violently. Prior to this violent failure one of three 3000 A main circuit breakers had tripped shutting off power to a large portion of the facility. The owner's maintenance electricians had tried to reset this breaker serving DS-2, but it had tripped out immediately with an electrical arc flash occurring at a tie bus joint, point F1 as shown in Fig. 6.2 and 6.3. Three maintenance workers were injured by the flying metal particles and superheated gas from this overhead fault. At this time the newspaper contacted the electrical contractor, who had originally installed the system approximately ten years prior to this incident, to provide troubleshooting service and to restore power to the facility.

The contractor's electrician arrived at the site and determined that he could probably restore power to the facility by disconnecting the tie bus connection at the tee connection point A between DS-2 and DS-3 as shown in Fig. 6.2 and 6.3. As the electricians were working on removing the tie bus jumpers with DS-2 deenergized, several newspaper employees were standing 8 to 10 ft. from transformer T2 and nearly underneath the bus duct from T2 to DS-2. The lead electrician was talking to a city fireman, and they were standing in the doorway to DS-2 when they heard an explosion and a shower of sparks flashed across the area outside DS-2 toward T2.

Three employees of the newspaper were enveloped in the electrical fire ball that had erupted. One employee later died from the second and third degree burns sustained over 70% of his body. A second underwent years of skin grafts and surgery for his second and third degree burns over 50% of his body; this man never worked again. The third man was burned with second degree injuries over 40% of his body and third degree burns on one hand. He later lost a finger to these injuries. Four of the other employees required hospital treatment for their injuries.

The local utility was immediately notified and within minutes had disconnected the 4160 V circuits feeding the two 2500 kVA transformers at the facility. The city fire department and police immediately secured the area—even the electricians were not allowed to enter, and the arson squad investigated the scene.

After the electricians were allowed to enter the area and begin their work, the contractor's lead electrician discovered that a great deal of moisture had entered the tie bus duct at fault point F1. This moisture was identified by the electrician as a likely cause of the faulting to ground that had originally caused the main 3000 A circuit breaker to DS-2 to trip. It is important to note that the bus duct from DS-2 to DS-3 was carrying no current since the 3000 A breaker in DS-3 was normally open (N.O.).

This engineer was retained by the attorneys representing the newspaper in the wide ranging lawsuits that were subsequently

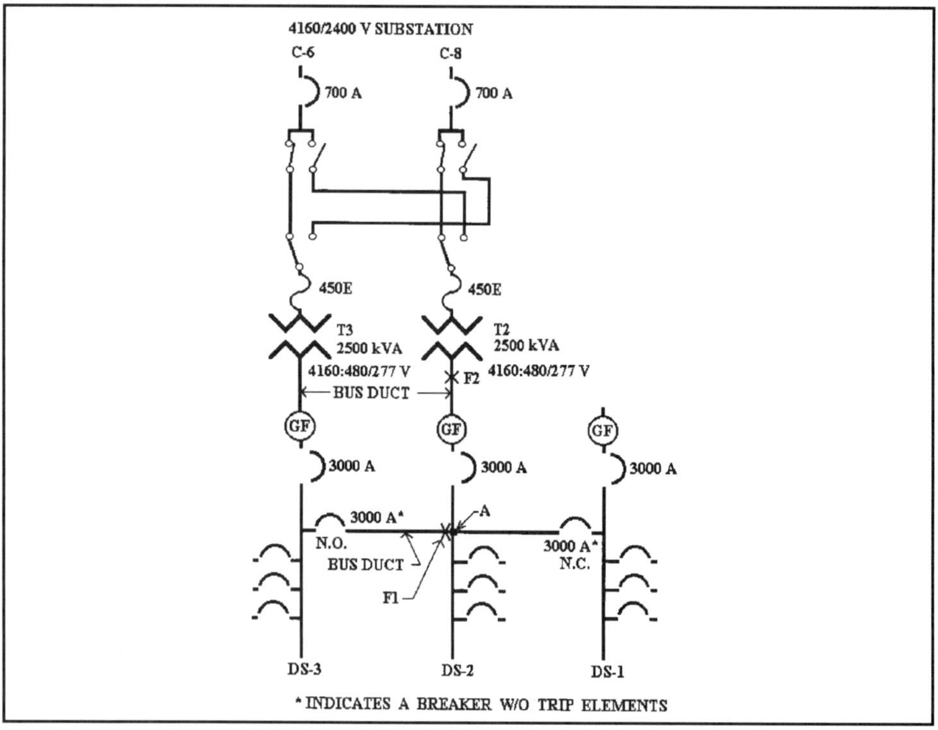

Fig. 6.2 *Bus duct case circuit.*

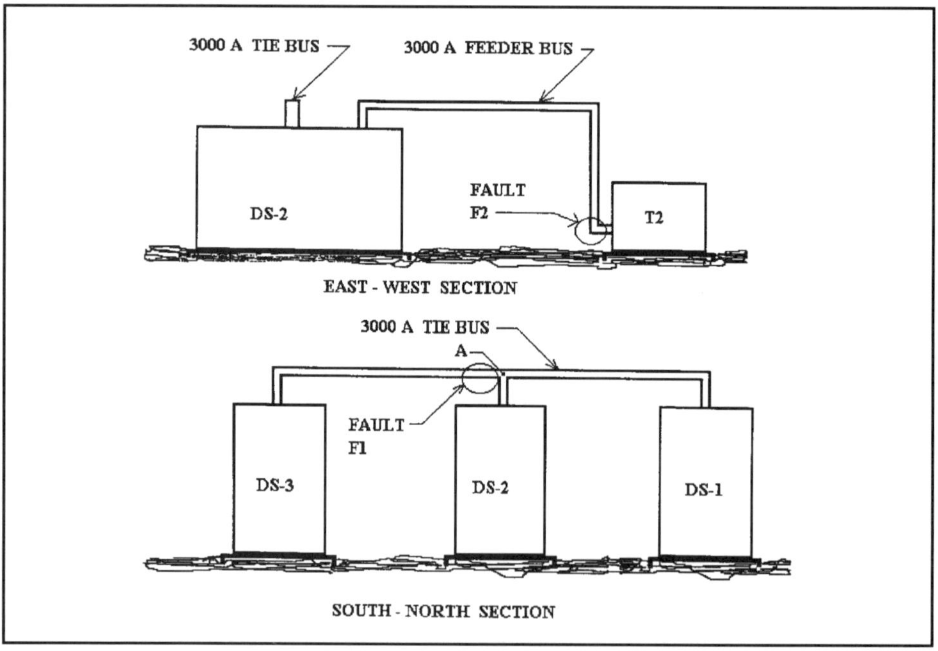

Fig. 6.3 *Electrical distribution center sections.*

filed. Nearly everyone who was involved in the design, construction, maintenance and electrical service of the facility was named as defendants in these multiple lawsuits.

The initial engineering investigation in this case consisted of the discovery of the basic sequence of events leading up to the explosion as has already been described. The weather in the area prior to the incident had been extremely rainy; in fact, rain had fallen nearly continuously during the 12 hr. period preceding the accident. Original electrical plans for the facility construction were obtained and detailed information was gathered regarding the protective devices, circuit breaker settings, available fault current, manufacturer's time-current curves for the fuses and circuit breakers, transformer impedances and ground fault protection. After approximately 4-1/2 hrs., the electricians were allowed to reenter the area following the explosion, and they proceeded to remove all faulted bus duct, test all bus duct intended to remain in place with a megger (high voltage tester for insulation integrity) and to install temporary cables to serve the facility loads. It was at this time that the lead electrician discovered the high moisture content of the tie bus duct near the tee point A in Fig. 6.2.

The available fault current from transformer T2 was estimated by the utility at 40,000 A. After the electricians had disconnected the tie bus connection at point A linking DS-2 to DS-3, and with the 3000 A main breaker (M.B.) in DS-2 open, the second fault F2 occurred. At this time the feeder bus from T2 to DS-2 was still energized but no load current had been flowing in this bus duct for about 45 min. since the DS-2 main breaker had last tripped.

A logical explanation for the second fault was apparent at this point in the investigation. The feeder bus duct to DS-2 had, prior to the DS-2 M.B. tripping off, been carrying the load of DS-2 and DS-1 which at that time of morning was about 800 Amps. This would cause the bus duct conductors to operate at a temperature of around 48° C (118° F) and would tend to drive moisture away from the conducting buses. When the load was disconnected, the bus bars would begin to cool down, approaching the outdoor ambient of about 32° C (90° F). As this cooling effect occurred, moisture would not only discontinue to be expelled, but would be drawn to the bus bars of the duct through any porous material nearby.

In November 1982, the feeder bus duct was opened by drilling the spot welds, careful chiseling, and metal sabre saw cutting of the sheet steel enclosure that houses the multiple aluminum bus bars which are the conductors. Fig. 6.4 shows the bus duct construction.

The feeder bus duct is rated by the manufacturer at 600 V, 3000 A, 3Ø 4 w, with full sized neutral, suitable for outdoor use having a totally-enclosed weatherproof, dust-resistant design.

The fault damaged section of bus duct showed clear evidence of moisture penetration with rust deposits on structural steel and insulating materials inside the duct. The urethane foam seal material used will readily absorb and retain moisture, and this material is only separated from the bus bars by two layers of polyester varnished glass cloth. The presence of moisture within the duct enclosure was concluded as the prime factor contributing to the fault at F2.

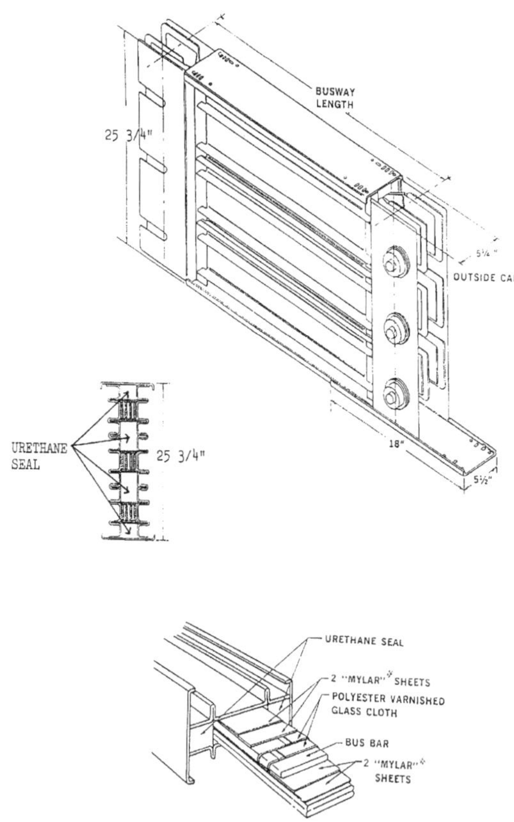

Fig. 6.4 *Bus duct construction.*

The secondary bus duct feeders from T2 and T3 were electrically protected only by the utility's overcurrent devices, the 450 E primary fuses and 700 A relaying circuit breakers. The 450 E fuses will allow a secondary fault of 11,440 Amps rms to exist for 16.67 min. before clearing the fault and the utility substation phase overcurrent relays will allow a current of 6,100 Amps rms in the secondary for at least 3.33 min. before clearing the fault. The electric utility found that after the second fault, two phase relays on C-8 activated which, along with the existing wye-wye transformer configuration, indicates that basically a line-to-line fault finally cleared the circuit. The extensive damage to the bus duct at F2 eventually involved all three phases and the neutral, however.

Arcing faults are known to frequently exhibit low current levels due to the apparent impedance of the arc itself. Tests on aluminum conductors within steel enclosures have shown that the instantaneous current values at various arc resistances from 0.01 to 0.2 ohms can vary from 200 to 15,000 amperes.[2] Obviously, the arcing can and frequently does persist at rms current levels insufficient to operate either circuit breakers or fuses. Meanwhile, the violent effects of the electrical arcing continue, creating temperatures of several thousand degrees Celsius with gas plasmas and explosions that scatter superheated metal particles which can destroy equipment and severely injure or kill bystanders.

In this case all experts agreed with the conclusions as to the cause of both faults F1 and F2 as previously given. The client was obviously concerned with the future reliability of the feeder bus duct that had not faulted from T3 to DS-3. Although this bus duct had been high voltage tested before the utility restored power to the facility, there remained a question of the efficacy of this bus duct to withstand future moisture penetration. It was the owner's decision, based upon the engineer's recommendation, to replace all outdoor bus duct with insulated cable in rain-tight wireway. In order to provide the 3000 A capacity of the feeder and tie bus ducts with cable in wireway, eight 500 kcmil 90° C rated copper (Cu) cables were required when derated to a maximum ambient of 45° C (113° F).*

* 1981 NEC Code rated ampacity for this cable configuration is 2,993 Amps.

The existing transformer secondaries to the buses were connected with 9 short sections of 500 kcmil 75° C rated Cu cable. In accordance with the 1981 NEC Code, these cables are rated at 380 A each for an ambient temperature not to exceed 30° C and derated to 82% of this ampacity for ambient temperatures not to exceed 45° C. The use of nine 500 kcmil Cu 75° C rated cables provides 2,804 Amps of capacity.

In order to better protect the multiple secondary paralleled feeder cables from T2 and T3 to DS-2 and DS-3, respectively, the engineer considered the following alternatives:

1. Installing ground fault interrupting circuit breakers in the padmounted transformer secondary compartments of T2 and T3.
2. Installing cable limiting in-line fuses on both ends of the secondary cables—8 limiters x 3 phases x 2 ends or 48 cable limiters per transformer and distribution center.
3. Installing primary current limiting fuses rated 400E at the transformer primaries to allow inrush currents to the transformer and to limit the maximum instantaneous peak current let-through.

The third alternative will limit the "i^2t" let-through to a secondary fault, but has the disadvantage of requiring at least 0.1 seconds fuse melting time at 12 times the continuous primary rating of the transformer (*i.e.*, 4,164 A). This is a continuous secondary current of 36,084 A rms, a value too large to expect a usual arcing fault in the secondary to activate the fuses.

The first alternative was not realizable due to the transformer secondary compartment space limitations and the multiple paralleled feeder cables planned.

This leaves the second alternative, which was ultimately installed using manufacturer recommended 500 kcmil copper cable limiters that will clear a 1500 A rms fault in 5 min. and a 3000 A rms fault in 10 sec. These potential arcing fault currents are considerably lower than the protection provided by the original system prior to the time of the accident. There is also an increased reliability associated with the main feeders to the facility's distribution centers in respect to moisture and dust induced failure.

Concurrent with this engineering design work to modify the electrical service at the newspaper, the law suit continued. Over two years after the accident, this engineer was still providing information regarding his investigation to attorneys for defendant and plaintiff in the matter. The engineer's deposition was taken in early April, 1983, and a trial was first set for late 1983. After several continuations of trial, all parties settled their law suits for an undisclosed amount. Knowledgeable sources in the case have estimated that the total settlement amount in these law suits was several millions of dollars.

6.09 A Case of EMF Concern for a New Power Line

In early 1982, an electric utility company requested a confirmation from the state Utility Commission of the state siting committee's prior approval for a two and one-half mile extension of the utility's 138 kV transmission line system to reach an existing substation presently served by a 46 kV subtransmission line. The confirmation was necessitated by an intervenor's motion for denial of the new line's siting which was filed on behalf of the residential property owners along the proposed route of the line.

This engineer was retained by the attorney representing the property owners to provide expert opinion and gather evidence in support of the owners' position. Crucial to the property owners' case was an already filed motion which cited the Wertheimer and Leeper study from Colorado (as referenced in Chapter 5, *Chapter Notes,* No. 56), and which epidemiologically showed a significant increase in childhood leukemia due to EMF exposure. Both litigants were well aware of this study by 1982.

The electric utility responded to this challenge to their planned expansion of facilities in the following way:

1. By citing the rules of civil procedure in the state which provide for relief from judgement because of the discovery of new evidence which by due diligence could not have been discovered in time to move for a new trial. This argument is generated by the original hearing proceedings, held in 1981, in which the property

owner's expert (not at that time this engineer) was unaware of the Wertheimer and Leeper (W-L) study.

2. By introducing affidavits from two experts: an electrical engineer who is DOE and utility oriented and a radiation biologist who was the program director for research in 60 Hz bioeffects at a heavily subsidized University by DOE and EPRI. Both experts criticized the W-L study and even went so far as to state that "[such work is] certainly not allowable in an acceptable scientific study."
3. By alleging that "the purported evidence is not newly discovered, is not credible . . . would serve no purpose other than to delay . . . is an attempt by the Intervenors to confuse the issues."

Despite the somewhat intimidating testimony of these utility experts, the engineering testimony for the property owners concentrated upon the methods of acceptable statistical analysis employed by Wertheimer and Leeper and their resultant conclusions. Unfortunately, the Swedish study published in June 1982, (as referenced in Chapter 5, *Chapter Notes,* No. 59), which confirmed the W-L work was not discovered by this engineer until after this hearing, but the large body of research that was known at this time solicited an engineering opinion which stated "there is . . . sufficient evidence of adverse biological effects due to power frequency electromagnetic fields on humans to suggest that we proceed with caution to transmit and distribute electrical energy . . . in close proximity to human habitation."

Further, this engineer criticized the close proximity of the 138 kV line to adjacent residential and institutional property and in particular the previously approved diagonal crossing of the transmission line over an elementary school recreational yard.

The state utility commission ultimately granted the electric utility's request, but a compromise was enacted regarding the line placement over the school yard and the line was constructed such that it remains on the opposite side of the two roadways near the school property and does not cross over the playground.

The subsequent EMF bioeffects research since 1982 has reinforced the engineer's opinions testified to at this hearing.

6.10 A Case of an Exploding Voltmeter

In May 1987 a 44 year old air conditioning mechanic was in the process of trouble-shooting a 200 ton centrifugal chiller located in the basement of a high rise government complex building. The chiller had shut down and the building maintenance engineer reported that the 400 A 3 pole circuit breaker (C.B.) protecting the 480 V feeder to the chiller would not reset.

Upon arriving at the job site, the a.c. mechanic located the tripped breaker, the chiller and its safety disconnect switch. He visually inspected the equipment and turned off the safety disconnect at the chiller. He had reason to believe that there was possibly a bad winding in the 150 hp motor driving the chiller, which he intended to check. First he returned to the motor control center where the C.B. for the chiller feeder was located and used his multimeter* to check the voltage at the load side of the breaker, which he found was zero (*i.e.*, the breaker was indeed tripped off).

Next he went back to the chiller, removed the covers to gain access to the motor electrical terminals and checked for grounded or shorted windings using the resistance measuring capability (ohmmeter function) of his instrument. He next checked the fuses and connections at the safety disconnect switch using both the voltmeter and ohmmeter functions of the multimeter. At this point he concluded that the motor was probably not at fault and that the safety disconnect and fuses were in good working order.

He then returned to the motor control center to check the line side of the C.B. for power (voltage). As he touched the first voltmeter probe to one of the line terminals, an electrical explosion occurred and he was knocked backward to the floor by the fire ball which came from the C.B. cubicle. His clothing was on fire and another workman in the area quickly extinguished this with a hand-held fire extinguisher.

The mechanic suffered third degree burns over 45% of his body which required extensive reconstructive surgery and skin grafting. He also suffered hearing loss in both ears and limited dexterity in

* A multimeter is an instrument that can be used to measure voltage, current, or resistance selectively.

his hands. After about 3-1/2 months he was able to return to limited work in the office of his employer. After a year he was able to return to working at his previous job as an a.c. mechanic.

The man had worked for some 13 years as an a.c. mechanic prior to this accident.

This engineer was contacted by the insurance carrier for the employer within a month of the accident and began an engineering investigation to determine cause. The accident site had been basically left as it was except that the electrical contractor, who had originally installed the motor control center and feeder to the chiller, had removed the C.B. involved and had retained it as evidence. The damaged multimeter was also available having been taken as evidence by the mechanic's employer. This engineer also owned an exact make and model of the subject multimeter which served as an exemplar in this investigation.

The cause of the explosion was concluded by the engineer in a report issued to the insurance company within two months of the initial site examination. It was stated therein that,

> it is most likely that the explosion . . . was the result of a phase-to-ground fault at the line side terminals of the . . . circuit breaker. This ground fault rapidly escalated into a phase-to-phase fault . . . and the resulting . . . explosion with expanding plasma gas and vaporized metal. . . . [It is] believed that the Ohm probe tip touched grounded metal, probably the breaker operating mechanism, while the left hand voltage probe touched an energized line terminal. This placed the resistance metering circuitry of the [omitted trade name] across a 277 V potential and caused a low resistance phase-to- ground fault. The Ohm probe is equipped with a 1 A 8AG glass body fuse rated at 250 V or less. This fuse can safely interrupt up to 35 A at 250 V, but will likely fail at voltages greater than 250 V and for fault currents greater than 35 A. This fuse obviously failed catastrophically.

The report continued as follows:

> At this time, I have no evidence to indicate that the breaker installation contributed to the severity of the

> accident or that it was a proximate cause for this accident. However, the use of washers to shim a 3/4" space in a bus connection is not acceptable practice in electrical work at these voltage and current levels. There also appears to have been a missing barrier with warning across the line terminals of the original breaker that was furnished by the manufacturer on the replacement breaker.
>
> The fuse protection for the MCC (motor control center) main disconnect and MCC feeder disconnect are not properly coordinated in that the closest upstream device to the fault should clear the fault first, which did not occur. This lack of coordination did not, however, cause the accident nor affect the accident severity.
>
> The catastrophic failure under the hypothesized energized phase-to-ground connection of the Ohm probe is a foreseeable event by the instrument manufacturer. In my opinion neither the design nor the warning issued with the instrument were adequate to protect the user from injury in this situation.

The engineer was deposed in May of 1989 and the case remained active until the early part of 1990 before a settlement was reached between the parties. The plaintiff mechanic was compensated by a $350,000 award for his injuries sustained as a result of this product failure.

6.11 A Case of Inadequate Illumination

In 1983 a female civilian food-service worker at a military base was assaulted while working. The worker had left the mess hall where she was employed at around 2 A.M. to carry some garbage to an outside disposal area. She was attacked by a man who "put a box cutter to her throat," cut and fondled her, and probably would have raped her if a passing car had not startled the attacker with its' headlights, frightening the man away.

A law suit was subsequently filed against the government for failing to provide adequate protection for this employee in the workplace. This engineer was contacted in 1985 by the worker's attorney to evaluate the adequacy of the lighting in the vicinity of the assault. The engineer visited the site at night and used a low level sensitive photometer to measure the illumination in the area of assault. General observations were made and recorded during this inspection regarding the surrounding light interfering structures and sources of illumination. The date of inspection was coordinated with the time of the moon to assure that moonlight was at a comparable intensity during the investigation.

The engineer issued a report to the client attorney stating the opinion that "the illumination of the trash enclosure area interior was deficient." Cited in this report were recommended standards of the IESNA for minimum levels of illuminance (footcandles) necessary to provide protective lighting and for safety. The measured illuminance behind the trash enclosure fence where the woman was attacked was less than 0.01 footcandles (FC). In addition, the facility was operated without light sources from the building to provide illumination within the trash enclosure. The only artificial light sources contributing to the illumination within the trash enclosure came from 1) a street light approximately 75 ft. diagonally N-W from the enclosure West wall and 2) the light emanating from the windows of a three story residence hall approximately 50 ft. North of the enclosure. The trash enclosure was surrounded on three sides by a seven foot high fence made of wire mesh with interwoven sheet metal visual barriers and on the South side by the North wall of the mess hall. Within the trash enclosure were several large metal trash boxes (dumpsters) of six foot height that offered dark hiding places.

The defense offered illuminance measurements taken with an instrument which the manufacturer stated was not intended to read less than 5 FC, and in a lighting survey report issued by the government's investigator, it was stated that "at the dumpsters, no light readings could be noted."

Regardless of these obvious deficiencies in the defense's case, the case was tried before a federal judge, who personally inspected the area at night, and who awarded the plaintiff $67,750 in damages from the military service responsible for the base operations.

6.12 A Case of a Misdiagnosed Electrocution

On July 4, 1974, during mid-morning a man weighing 214 lb. and wearing shoes, trousers, leather gloves, but no shirt, was using a hand-held electric circular saw to cut firewood at his residence.

The ground was wet as it had been irrigated the day before. He had been working for about two hours at the task when he was discovered lying dead on the ground face down. The saw was under his abdomen and continuing to run with the spring guard covering the rotating blade.

At the time of its operation, the saw was connected to an electrical outlet by two extension cords. One extension cord was a three wire cord with ground and the other cord was a two wire cord without ground. The three-prong to two wire adapter plug provided with the circular saw by the manufacturer was not used by the decedent.

The fire and police department investigators who responded to the accident requested a City Electrician to be called to the scene to examine the electrical components for possible malfunction that could have electrocuted the man. The electrician's investigation found that the three wire extension cord had been damaged and that there were some frayed exposed conductors visible in the cord when the insulation tape used to repair the cord was removed.

Upon completion of his resistance and continuity checks, the electrician turned on the saw and demonstrated to the other investigators that he could hold the saw in his hand and place the other hand firmly on the wet ground without being shocked. He found no indication of any voltage on the case or handle of the saw. An ammeter attached to the saw and cord indicated no abnormal current flow.

The saw, extension cords, and leather gloves were sent collectively to three different engineering testing laboratories at different times over the next six months for evaluation. The laboratory tests all indicated that:

1. The case of the saw was effectively electrically connected to the ground wire and prong of the saw cord and male plug.

2. The hot and neutral legs of the saw cord were isolated electrically from the saw case and ground wire by over 2,000,000 ohms of resistance, respectively, which is normal for this product and safe for the intended use.

3. Leakage current was measured according to the UL 45 and ANSI C101.1 test procedures and standards. The

maximum reading obtained was 0.04 mA. The maximum allowed by UL and ANSI is 0.5 mA. The manufacturer's specification is 0.2 mA max.

4. The saw was subjected to running tests to heat the insulation and 24 hr. humidity tests to introduce moisture into the insulation system. The leakage current tests were repeated, with maximum leakage current measured at 0.08 mA and no breakdown of dielectric strength.

5. By using the decedent's gloves, it was found that the size and thickness of these gloves resulted, in many positions of operation, in holding the trigger switch closed even when no pressure was applied by the index finger.

6. Upon removal of the insulation tape around the extension cord at two locations, a lateral cut through the insulation of all 3 conductors was found. The tape was old, probably installed before the incident. The method of reinsulating the extension cord was not such that it would provide a reliable and safe use of the cord since it did not reinsulate the individual conductors adequately. With abrasion and flexure, the cord could fail electrically.

Shortly after the accident, the County Medical Examiner performed a postmortem examination and concluded the cause of death was electrocution. In the medical examiner's opinion, the burns on the decedent were electrical rather than thermal. He opined that electrical current had entered the body from the saw through a 3 inch diameter burned area on the decedent's mid-abdomen and that this would likely have produced a cardiac arrhythmia or fibrillation of the heart. The examiner also found a small elongated burn on the back of the right hand toward the thumb over the bones of the hand that go to the index and middle finger.

A law suit was brought in this case for products liability against the saw manufacturer and the 3 wire extension cord manufacturer. It was alleged that the saw was defective in design because it was not double insulated and that the manufacturer failed to provide adequate warnings in the instructions which accompanied the saw as to the danger of using the product with an ungrounded electrical supply or two wire extension cord.

Against the 3 wire extension cord manufacturer, it was alleged that the cord was defective in design and manufacture because of the cuts in the insulation that had exposed bare conductor wires and because of inadequate warnings provided with the cord concerning the danger of using the cord with such insulation damage.

The trial judge in this case granted summary judgement in favor of all defendants, and the judgement was appealed. On appeal, the court held in favor of the summary judgement for the cord manufacturer, but reversed the judgement in favor of the saw manufacturer.

In the court's opinion it was stated that genuine issues of material fact existed as to defectiveness of the design of the saw, adequacy of warnings given in the instructions accompanying the saw, and as to whether the use of an ungrounded extension cord was an unforeseeable misuse of the product. The appellate court's decision was dissented by one judge who agreed with the trial court's decision in favor of the cord manufacturer, but disagreed in the reversal against the saw manufacturer. This judge's opinion was that the express instructions and warnings furnished by the saw manufacturer were unmistakably clear as to the consequences of using the saw with a two-prong extension cord. "A skull and crossbones was not required." The judge further said "since it is clear that the saw would not have been energized with an electric current if a grounded extension cord had been used, the saw was not unreasonably dangerous as a matter of law, regardless of whether there was an alternate design (double insulation) available."

This engineer was retained in January 1985 to advise the defendant saw manufacturer's attorney in regard to the scheduled trial of the case and to testify at trial if warranted. The engineer was furnished with the first trial depositions of three engineers and the county medical examiner, and with various reports of testing performed on the subject saw and cords. In addition he was provided with the pleadings to the court of appeals, the decision of the court, the instructions for operation furnished with the saw and an engineering evaluation of the case by Ralph H. Lee.*

---

* Lee Electrical Engineering, Inc. Oct. 18, 1979. Mr. Lee, who is deceased, was a recognized consulting electrical engineer, particularly in the field of lightning protection. He served as Chairman of the NESC Committee, ANSI C2-1981 and 1984.

In Lee's report, he made various estimates of the resistance of the electrical circuit through the victim in order to determine a range of reasonably probable shock current in this case. His estimates varied from 6 mA to 80 mA under differing conditions of gloves-no gloves, shoes-no shoes, and one hand-two hand contacts. He also estimated the fibrillation current for this 214 lb. man at 95 mA for 0.5% of all individuals (*i.e.*, greater than 95 mA for 99.5% of all individuals of this weight).* Lee concluded that "the evidence shows that the victim died from cause [sic] other than electrocution."

This engineer's evaluation of the evidence supports this conclusion. The medical examiner was in error when he concluded that the burn was electrical because "the appearance of the burn . . . in the center . . . had the characteristic . . . dried desiccated appearance of an electrical burn as opposed to a thermo [THERMAL] burn." When asked in his deposition, "How would the appearance of that burn been different from a thermo [sic] burn if a hot object was held against the abdomen for a period of time?" the examiner answered "It would vary and depend somewhat on the intensity of a thermo [sic] burn. The most remarkable difference is a marked reddening in the case of a thermo burn possibly with a blood formation or blistering and as opposed to the dry type of burn that you get with electricity in which there is no reddening or blistering." The question was then asked "Are you saying there was no redness and blistering?" to which the answer was "Not in the center of it (a 3 inch diameter burn area) that I referred to at the margins. As I said, there was probably some thermo [sic] burn."

When asked how deep the burn was, the medical examiner had no notes on that aspect and didn't recall.

Mazer states in his three volume handbook on electrical accident investigation that deep subdermal injury is usually more closely associated with electrical burns than thermal burns and that thermal burns damage the skin more severely.[3] He further states that it is estimated that the lowest current that can cause a noticeable first or second degree burn in a small area is about 100 mA flowing for about one second[4] (as has already been stated in a study cited

---

* See Chapter 5, p. 77.

in Chapter 5, p. 78, for electrocutions by contact at less than 1000 volts, 45% had no perceptible electrical burns).

In respect to the possibility expressed by the medical examiner that the abdomen burn area and right hand burn area represent entrance and exit wounds for electrical contact, this is impossible if the saw case was energized as suggested since it would be at a constant potential (with respect to ground) and there would be no electrical force to cause current to flow between abdomen and hand.

These observations, along with the engineer's opinions regarding the efficacy of double insulated hand tools vs a third wire ground were communicated in a report to the client attorney prior to trial.

Rather than try the case, the manufacturer decided to settle by "payment of a nominal amount to plaintiff" and the file was closed in this matter.

Chapter Notes

1. *1990 NESC*, IEEE. p. 47.

2. Richard R. Conrad and Derio Dalasta, "A New Ground Fault Protective System for Electrical Distribution Circuits," *IEEE Trans, on Industry and General Applications,* May/June 1967. p. 222.

3. William M. Mazer, D.E.E., P.E., *Electrical Accident Investigation Handbook,* Electrodata Inc., Glen Echo, MD. 7.4.1.0.

4. Ibid, 3. 7.4.2.0.

About the Author

Robert E. Nabours, Ph.D., P.E. has been involved in the construction industry for more than 40 years.

During the early 1960s, he was an instructor in electrical engineering at the University of Arizona, a research engineer in advanced communications systems, and the chief engineer for an Arizona electronics manufacturer. Since 1968, he has been engaged full time in the private practice of design electrical engineering, serving clients in the construction industry and providing forensic electrical engineering services to the legal profession.

He received his Bachelor of Science in Electrical Engineering from the University of Arizona in 1957, his Master of Science in Electrical Engineering from Stanford University in 1959, and his Doctor of Philosophy in Electrical Engineering from the University of Arizona in 1965. He is a registered professional engineer in Arizona, California and New Mexico and is a member of the National Society of Professional Engineers, a senior member of the Institute of Electrical and Electronics Engineers, and is a Fellow of the National Academy of Forensic Engineers. He is also a member of the technical honorary societies of Sigma Pi Sigma (Physics) and Tau Beta Pi (Engineering).

He has been married to his wife, Jane, for over 40 years. They have three children and six grandchildren all who live in Arizona. He and his wife are students of Western U.S. history, and avid collectors of Southwestern Indian art. They love to travel both as a couple and with their family, and he enjoys golf, hunting, fishing and wilderness exploring with his sons and grandchildren.

Index